Shortwave Radio Listening for Beginners

To all the people that helped to make
my more than 20 years of long-distance radio listening so enjoyable!

Shortwave Radio Listening for Beginners

Anita Louise McCormick

TAB Books
Division of McGraw-Hill, Inc.
Blue Ridge Summit, PA 17294-0850

Trademarks

Bearcat®	Uniden Corporation
Drake®	R. L. Drake Co.
IBM®	International Business Machines, Inc.
Icom™	Icom, Inc., Osaka, Japan
MicroDec™	Somerset Electronics, Inc.
Radio Shack®	Tandy Corporation
Sangean™	Sangean America, Inc.

FIRST EDITION
SECOND PRINTING

Library of Congress Cataloging-in-Publication Data

McCormick, Anita Louise.
Shortwave radio listening for beginners / by Anita Louise McCormick.
p. cm.
Includes index.
ISBN 0-8306-4136-X ISBN 0-8306-4135-1 (pbk.)
1. Shortwave radio—Amateurs' manuals. 2. Shortwave radio–Receivers and reception. I. Title.
TK9956.M378 1993
621.3841'6—dc20 92-41603
CIP

Acquisitions Editor: Roland Phelps
Editorial team: Peter D. Sandler, Editor
Lori Flaherty, Managing Editor
Production team: Katherine G. Brown, Director
Wanda S. Ditch, Layout
Ollie Harmon, Typesetting
Susan E. Hansford, Typesetting
Nancy Mickley, Proofreading
Design team: Jaclyn J. Boone, Designer
Brian Allison, Associate Designer

EL1

Contents

Foreword

LONG BEFORE ELECTRONIC COMMUNICATIONS WERE INVENTED, PEOPLE HAD A strong desire to communicate with each other. The desire to convey information to each other, whether through a simple greeting, a story of something fantastic, or a dire warning of impending danger, is instinctive.

In the early part of this century, a means of sending information and ideas by wireless methods was invented, and this gradually developed into modern radio as we know it today. The urge to disseminate information, tell stories, and entertain each other is very much present in the tens of thousands of radio broadcast stations around the world. There is scarcely a place on earth where you can't hear all manner of radio programs with simple and inexpensive radio equipment.

Strangely, despite the ever-increasing number of stations, a type of homogeny has occurred. Perhaps it is something about our times. One can travel from coast to coast across this great land and hear radio stations with similar formats. But there *is* a place in the radio spectrum where all is not so predictable.

The international shortwave radio bands are a much-needed alternative source of news, information, music, and commentary—all of these presented from the unique perspective of the source country.

Shortwave Radio Listening for Beginners can help open the door to the many strange and wonderful things available only on the shortwave band. In this restless and complicated world, shortwave listeners can broaden their horizons by discovering the many alternative viewpoints offered on shortwave radio. With a simple receiver, distant lands can be touched, different opinions heard.

Anita Louise McCormick's well-researched book on the shortwave experience will guide you through this exciting, alternative territory. It also introduces you to the world of long-distance AM listening, ham radio, and scanning. Listen, learn, and have great fun along the way!

Allan Weiner, President
Radio New York International

Acknowledgments

I WOULD LIKE TO THANK ALL THE PEOPLE, COMPANIES, AND ORGANIZATIONS that helped me prepare this book by supplying photos, information, and support—including Radio Shack, Sangean, Grundig, Drake, Icom, Grove Enterprises, Radio New York International, Radio Free New York, the A*C*E*, the Bearcat Scanner Club, the American Radio Relay League, the National Amateur Radio Club, WB2JKJ, and many others.

And special thanks to the people at TAB Books for helping me get my first book into publication!

Introduction

IF EVER THERE WAS A HOBBY THAT HAS SOMETHING FOR EVERYONE, IT'S LONG-distance radio listening.

- Do you like music, ball games, news, talk shows *you* can participate in, religious programs, old time radio comedies, westerns, and dramas from the thirties, forties, and fifties? If so, the AM band has plenty to offer.
- Want to tune in on police action, fire department calls, utilities, aviation air-to-ground communications, and countless other as-it-happens transmissions? Then you'll love the scanner action bands.
- Perhaps your tastes run to the international. For an investment of as little as $100—sometimes even less—you can own a shortwave radio that brings programs from all over the world into your home. Imagine how much fun it would be to switch on your world band radio and hear broadcasts from distant lands such as Israel, Brazil, Australia, Germany, Japan, South Africa, Holland, England, Czechoslovakia . . . even Russia, the Ukraine, and many of the other countries that were once part of the Union of Soviet Socialist Republics.

And you don't have to learn a foreign language to understand what they're saying. Dozens of countries beam programs in English to their fast-growing North American audiences every day, especially during the evening hours when we are most likely to tune in.

- If you think you'd like to do more than just listen, how about starting a station? As a ham radio operator, you can communicate with other radio enthusiasts from across the country and around the world.

Morse code has been one of the biggest stumbling blocks to new ham operators for decades. But the Federal Communications Commission recently made some positive changes in their regulations. A No-Code Technician class license is now available that gives you voice privileges on the high-frequency ham bands. And Novices, who had only been allowed to operate in radiotelegraph (Morse code), are now permitted to use voice on certain portions of the ham bands.

Once you've passed the FCC exam—and 5,000 new hams are passing it in the United States every month—you can set up a transmitter and put a signal of your own out over the airwaves.

Welcome to the exciting and ever-changing world of radio! Long distance radio listening has been popular in Europe, Asia, and many other parts of the world for decades. And now it is starting to catch on big here in America. With the help of this book, *you* can be a part of it.

Chapter 1 explores the history of radio. Starting with Marconi's early experiments, it takes you through the golden years of radio—right up to where radio is today. You'll see how broadcasting changed and grew through the decades from a scientist's lab experiment to a multimillion-dollar, worldwide industry.

Chapter 2 tells you about long-distance AM listening—an activity you can enjoy by using any AM radio you already own. Nearly everyone is surprised to discover that you can hear stations several hundred miles away by simply turning up the volume and tuning S-L-O-W-L-Y between local stations.

Try it! You'll be surprised what you can hear. The only hard part is that you'll have to wait until evening. The way our atmosphere is designed, AM signals aren't able to travel very far when the sun is up, but as soon as it dips below the horizon, watch out! There will be more signals than you ever knew existed, zapping in until morning.

Chapter 3 helps you discover how easy it is to hear broadcasts from overseas on shortwave radio. Daily programs of news, editorials, music, features, mailbag shows, and contests you really can win—it's all waiting for you to tune in and listen! With all the fast-paced changes going on in the world, you'll be sure to find plenty of excitement here!

Chapter 4 profiles some of the stations you're likely to hear on the shortwave bands, including: The British Broadcasting Corporation, Radio Moscow, Radio New Zealand International, HCJB—Ecuador, Radio Nederland, Radio Czechoslovakia, Radio for Peace International—Costa Rica, WWCR—Nashville, Tennessee, and the Voice of America.

Once you've had a chance to experience long-distance listening and see how fascinating it can be, you're bound to have some questions. In chapter 5, you'll learn the meaning of the most common radio abbreviations and terms. You'll also learn how to write reception reports that are useful to foreign stations—and receive attractive souvenir cards in return.

Chapter 6 covers the "action bands" and scanners. You'll also learn about the conditions that make long-distance UHF/VHF radio and TV reception possible.

Chapter 7 tells what ham radio is and how to go about applying for your license. Ham radio basics are explored. And you can read about some well-known operators you might meet on the air once you get your ticket.

The transmissions you'll hear as a long-distance radio listener have traveled hundreds, if not thousands, of miles to reach your receiver. Chapter 8 gives you a rundown of how radio waves travel from one place to another—in language anyone can understand. You'll see how sunspots, geomagnetic storms, seasonal changes, and even the time of day affect the signals you want to hear.

And, at the end of the book, is a list of resources for radio magazines, listeners' clubs, and mail-order catalogs that can help you get even more enjoyment out of your listening activities.

Unlike other books on radio hobbies, *Shortwave Radio Listening for Beginners* is written for people who would like to find out what they can hear, and start listening right away without having to wade through tons of technical information they don't really need as a beginner.You don't have to be a radio engineer to hear and enjoy foreign broadcasts any more than you need to be a television technician to watch TV. In fact, today's modern radios make long-distance listening as easy as operating your stereo or VCR. And it's a lot more fun.

When you get more involved with your hobby and discover which bands, stations, and activities you like best, you'll probably want to purchase more detailed books or join clubs that specialize in the areas you find most interesting. But for now, you hold in your hands all the information you need to tune in the most exciting broadcasts the world has to offer.

Have fun!

1
How did it start?

WITHOUT A DOUBT, THE FIRST LONG-DISTANCE RADIO LISTENER WAS GUGLIELMO Marconi (Fig. 1-1). Marconi had the honor of hearing the first radio signals to ever cross the Atlantic Ocean. But before he could accomplish that, he had quite a task ahead of him. He had to come up with a way to transmit radio signals and receive them at greater distances than anyone dreamed was possible.

Marconi—a pioneer of radio

As a boy, Guglielmo Marconi had always been interested in science. He enjoyed talking to professors when they came to his father's house to visit. And when he was sixteen years old, he built his first electromagnetic (radio) wave transmitter.

By the time Marconi started his research in the late 1800s, radio was already in its early stages of development. The German physicist Heinrich Hertz had recently invented the spark-gap exciter, a battery-powered device that could send a spark across a small space of air between two ball-shaped electrodes and, at the same time, produce a similar spark on a loop antenna several feet away.

Since the mid-1880s, telegraph operators had been sending their "dit-dah" messages in Morse code across the country. The messages traveled through thin metal wires in the form of electrical impulses. Hertz went one step further. He proved that electrical energy didn't necessarily have to be confined to a wire but could be transmitted through small gaps of air as well.

Marconi was inspired by Hertz's idea and used it as a basis for his own research. His goal was to find a method of transmitting these electrical impulses over greater and greater distances so they could be used not only for laboratory experiments, but for long-range, "wireless" communication.

With the encouragement of his mother, Guglielmo Marconi took on the world of technology and attempted to do what scientists many times his age had not been able to accomplish. "Guglielmo's mother was, as always, his chief aide in time of crisis.

1-1 Guglielmo Marconi—the first long-distance radio listener.

She understood that he must have a laboratory and she gave him the run of the top floor of the house."*

But his father's attitude was just the opposite. He was upset at his son's "foolish" ideas and yelled at his wife for permitting Guglielmo to waste time on such "nonsense."

"Giuseppe protested furiously at the way his son was employing every waking hour. He mercilessly attacked Annie for having allowed her son to waste irreplaceable years Guglielmo had dallied away in his youth—and whose fault was it? Who encouraged him?"*

But even though his home environment was not all that it might have been, Guglielmo Marconi refused to be discouraged. Marconi's early transmitting devices were able to broadcast waves of electromagnetic energy from one end of the room to another. And for a time, it was a mystery to him exactly why this was happening. But once he discovered the principles that made it work, he knew that he was onto something important.

"My chief trouble," he said, "was that the idea was so elementary, so simple in logic, that it seemed difficult to believe no one else had thought of putting to it into practice."*

By experimenting with various materials and antenna arrangements, Marconi found ways to gradually increase the distance his radio waves could travel. When he managed to get a signal all the way from his room to the end of the family garden (about 30 feet away), he finally convinced his father that he was onto something worthwhile.

---

*Marconi, Degna. 1962. *My Father, Marconi.* New York, Toronto, London: McGraw-Hill Book Company, Inc. (pp. 21-23).

Of course, Marconi was pleased to finally receive his father's support. But he knew that he had a long way to go—that his radio waves would have to cover much greater distances and make communications possible across natural obstacles, such as oceans and mountains—before the rest of the world would see the value of his invention.

By the time he was twenty years old, Marconi was broadcasting his radio signals over a distance of a mile and a half. But the materials he needed for research were getting more and more expensive, so he applied to Italy's Ministry of Posts and Telegraphs to obtain funds to continue his experiments. Unfortunately, they saw no value in his work and turned down his request.

Marconi packed up his bags and took his "black box" transmitter to England to see if their government would be interested in assisting him. Britain had a large navy and could certainly make use of such a device for ship-to-shore communications. But almost as soon as he arrived, disaster struck. His black box was confiscated by British inspectors who thought it might contain a bomb and decided that the best course of action was to destroy it. A relative helped him rebuild his invention, then took him to a patent lawyer. After months of endless paperwork, his transmitting device was finally registered.

During the next four years, Marconi kept himself busy perfecting his inventions and finding new ways to demonstrate their usefulness in public. In 1899, he made England's royal family happy by setting up radio communications between land and the royal yacht.

But all the while, Marconi dreamt of his *big* experiment—the day he would attempt to build a transmitter that could send radio waves across the vast expanse of the Atlantic Ocean. He knew that the equipment required to generate such a powerful signal would have to be at least 100 times stronger than anything he had built or used so far. The antenna would have to be exactly right, and so would the transmission and receiving sites.

Marconi installed 200-foot-tall antenna towers for his experiment at Cornwall, England. But before he had a chance to use them, a cyclone blew in and destroyed everything. Instead of trying to duplicate the original design, which would take more time and money than Marconi could afford, he decided to try a simpler design and see if it would work. He used two 150-foot poles with copper wires strung between them. While the original towers had been in the works for almost a year, the new antenna design took only two months to complete.

Next, Marconi looked to America to set up his receiving station. Towers were constructed at Cape Cod, Massachusetts. But again, the weather turned against him. A storm blew in and the whole project was in ruins.

But still, he did not give up.

Marconi left Liverpool, England, and set out for Canada by ocean liner. He then arranged a meeting with Newfoundland's governor to discuss how wireless communication could help to prevent loss of life at sea. The governor was pleased to hear about Marconi's invention and offered him assistance, along with temporary use of land to pursue his work.

After studying a map of Newfoundland, Marconi chose Signal Hill in St. John's for the receiving site. This time, Marconi had a totally different approach, one he was certain would work. Instead of building another set of towers for the next storm to take down, he decided to use the wind at this gusty seaport town to his advantage. He would raise the antenna wire with kites or balloons. Just one balloon—with a diameter of 14 feet—could hold 1,000 cubic feet of hydrogen and lift up to 10 pounds of antenna wire in the air.

With the government on his side and no antenna tower to collapse, it looked as if nothing could go wrong. But it did. When Marconi was testing one of his balloons on the morning of his big experiment, an unexpected gust of high wind broke the rope and the balloon was lost at sea. As he always had in the past, the undaunted Guglielmo Marconi went on with his work, using whatever equipment remained available to him.

The time of the experiment was fast approaching. At 12:30 P.M., his friend in Cornwall, England, would be sending the first transmission. The whole world was waiting to see what would happen. No one, not even Marconi knew for sure how radio waves would behave over such incredible distances. Would they curve around the earth, as Marconi expected—or would they travel in a straight line and be lost somewhere out in space?

Marconi selected a kite and took it outside to raise his antenna. Even in gale force winds and a downpour of icy rain, the kite flew boldly up into the sky. It soared courageously, going higher and higher until it was more than 600 feet above the ground.

Finally, the moment he had been waiting for arrived. The message was sent from England, and the first letter of the transmission, the letter "S" (three short clicks in Morse code), crossed the Atlantic Ocean.

Marconi heard it. And, at the age of 27, he became the world's first long-distance radio listener by monitoring a signal that had traveled farther than 2,000 miles to reach its destination!

Two days later, the experiment was attempted again, but failed on account of bad weather. Nevertheless, history had been made. And the world of communication would never be the same.

Now that it had been proven that radio waves could cross distances as great as the Atlantic Ocean, the scientific community was more anxious than ever to understand the principles that made long-distance radio communication possible.

A. E. Kennelly and O. Heaviside came up with the theory that radio waves were somehow bent by the upper layers of the atmosphere and returned to earth, making it possible to hear broadcasts hundreds, if not thousands, of miles away from the transmission site. These electrically charged layers of the atmosphere, which we now know as the ionosphere, acted as a type of "radio mirror" and made Marconi's experiment a success.

Businessmen were interested in cashing in on the benefits this amazing new wireless telegraph system offered. They built high-powered transmitters and constructed gigantic antenna towers on both sides of the Atlantic to send and receive messages. Letters transported by boat took weeks, sometimes even months, to arrive. But wireless messages zapped across the ocean at the speed of light!

Marconi started a station at Cape Cod and charged 50 cents a word to transmit messages to Europe. But while wireless had the advantage of speed, there was one drawback. Privacy was sacrificed. Anyone that owned a radio receiver could listen in.

For a time, it seemed that the wireless would be limited to military use, ship-to-shore communications, and transmission of overseas messages that the sender didn't mind sharing with the public.

But more discoveries were yet to come.

Once experimenters found a way to transmit voice and music over the air, wireless took on an entirely new direction. People from all walks of life who had never been interested in the "dit-dah" Morse code transmissions now wanted to own receiving sets. This discovery was more than a breakthrough for scientists; it was the birth of a whole new industry.

The early years of broadcasting

Ham radio operators were the first radio broadcasters. In the early years of radio, technology was still in the experimental stage, and everyone who could come up with a way to play music over the air quickly gained an audience. In fact, many early commercial stations were started by ham radio operators. They set up stations under their own call letters and later transferred them to the company that employed them. As a consequence, it is difficult to tell exactly when they went from being a hobbyist to a full-fledged professional broadcaster. (Unfortunately, ham radio operators are no longer permitted by the FCC to air music or programming intended for a general listening audience.)

Electronic equipment manufacturers financed a number of early radio stations, mostly to create public interest in their products and generate sales for radios. By today's standards, the first commercially manufactured radio receivers were poorly designed and difficult to operate. They had no way of distinguishing between nearby stations and often brought one program in on top of another. Still, people wanted to own a radio. They were fascinated with the whole idea of wireless communication.

The first commercial stations

Three of the earliest stations to broadcast regularly scheduled programs to the public were CFCX in Montreal, Quebec; WWJ in Detroit, Michigan; and KDKA in Pittsburgh, Pennsylvania. In 1919, CFCX decided to take to the air with weather reports and programs of recorded music. At the time, their only audience was a widely scattered group of amateur radio operators and radio men on ships at sea. They broadcast the first-known program of live music in 1920.

WWJ in Detroit, Michigan, officially began airing transmissions on August 20, 1920. It didn't take them long to find an audience. Letters poured in from excited listeners everywhere. Some people with exceptionally good receivers had even been able to pick up their signal over a hundred miles away!

As news of these "wireless broadcast stations" got around through word of mouth, newspapers, and magazines, everyone wanted to own a crystal set—the first radio available to the public at a price that nearly everyone could afford.

Crystal sets were made up of only a wire antenna, earphones, and a few inexpensive electronic parts. They had no way of dividing between stations that operated on nearby frequencies or of amplifying weak signals from out-of-town broadcasters as today's receivers can. But to the proud owners, they were an ultramodern, almost miraculous piece of equipment that made it possible to hear voices and music transmitted by people in studios miles away!

Another early broadcast station, KDKA, was started by Dr. Frank Conrad, ham radio operator, Westinghouse engineer, and scientist (Fig. 1-2). KDKA assembled its original studio on top of the Westinghouse building in Pittsburgh, PA. The station aired its first scheduled broadcast, the Harding-Cox election results (Fig. 1-3), on November 2, 1920.

THE FATHER OF RADIO BROADCASTING, Dr. Frank Conrad, Westinghouse engineer and scientist, whose experimental broadcasts led to the establishment of KDKA Radio, Pittsburgh ,and modern radio broadcasting.

1-2 Dr. Frank Conrad. Radio KDKA.

KDKA is credited with many firsts in broadcasting. On January 2, 1921, the first regularly scheduled church service was aired from Pittsburgh's Calvary Episcopal Church. During the same month, another first took place. KDKA hired Harold W. Arlin (Fig. 1-4), an electrical engineer, as the first full-time announcer. He introduced a number of celebrities, including Babe Ruth, Herbert Hoover, and Will Rogers, to his listening audience. And he did the first on-the-air, play-by-play broadcast of a ball game.

RADIO'S FIRST BROADCAST occurred November 2, 1920 on KDKA Radio, Pittsburgh, with the Harding-Cox presidential election return coverage.

1-3 Harding-Cox election. Radio KDKA.

Naturally, KDKA wanted to do everything possible to get their signal out, increase their broadcasting range, and of course, gain more listeners. In one experiment, they wired an antenna to a dirigible bearing their station letters and flew it high above the city (Fig. 1-5).

Westinghouse was so pleased with the success of KDKA that they made plans to construct three more broadcast stations. And Westinghouse is still in the broadcasting business. If you live in the eastern part of the United States or Canada, you can probably hear KDKA 1020 in Pittsburgh, Pennsylvania; WBZ 1030 in Boston, Massachusetts; and KYW 1060 in Philadelphia, Pennsylvania on your AM dial–all of which were started by Westinghouse in 1921.

Of course, the original equipment is no longer in operation. All three stations now use 50,000-watt state-of-the-art transmitters—and put out signals you just can't miss.

Everyone wants a radio

If you lived in the Pittsburgh area in the early 1920s and wanted a radio that brought in more stations than a simple crystal set was able to pick up, you could buy one from

RADIO'S FIRST FULL-TIME ANNOUNCER was KDKA Radio's Harold W. Arlin, shown here in a 1920's publicity photo. Arlin handled the first play-by-play of baseball and football, and introduced many noted persons in their radio debuts on this Pittsburgh station.

1-4 Harold W. Arlin. Radio KDKA.

Westinghouse for $10, provided you were lucky enough to get to the store before they were all sold out.

Radios took more time to manufacture than they did to sell—and some people

AN EXPERIMENTAL ANTENNA carried aboard a dirigible is tested in the Pittsburgh area by KDKA Radio personnel. Photo taken in the early 1920's.

1-5 A dirigible lifted KDKA's antenna high in the air. Radio KDKA.

didn't want to wait. Instead of counting the days until the next shipment came in, they went out in search of a ham radio operator and employed him to build them a radio from whatever spare parts he had available. When the word got out, even some "garage radio factories" received orders faster than they could fill them.

Westinghouse was excited by their success. They went right to work on their next entry into the market—a better-quality radio priced at $125. This turned out to be far too much for the average consumer to pay for entertainment at the time, so instead of selling out as the previous models had, the new model ended up gathering dust on store shelves.

In the summer of 1921, a small but improved crystal radio set with a reception range of about 15 miles, known as the *Aeriola Jr.*, was offered to the public for $25. It came complete with an antenna, headphones, and operating instructions. Even though $25 was still a lot of money to spend on entertainment, people flocked to the stores. They wanted to own receivers that could bring news, music, and an ever-growing variety of exciting programs into their homes.

Improvements in electronic parts and circuit design brought even better, easier-to-operate sets to the marketplace. Both Westinghouse and General Electric built tube-powered radios that could pick up weak signals earlier models hadn't been able to detect. Then loudspeakers came onto the scene, and instead of having to take turns with headphones, the entire family could listen in at the same time.

Radio stations everywhere

In the early 1920s, the radio industry grew at an astounding rate. Hundreds of broadcast stations took to the air. And when "skip" came in, listeners were thrilled to pull in transmissions from stations halfway across the country.

During radio's early years, programming often consisted of no more than an hour or two of music after supper. But people wanted to hear more, and stations happily complied with their wishes, expanding their schedules to include regular newscasts, weather, political speeches, children's programs (Fig. 1-6), farm reports, church services, ball games, and stock market reports.

People in rural areas had often felt cut off from the rest of society. But now they

1-6 An early children's program. Radio KDKA.

were thrilled to be able to turn on their radio and hear what was going on in neighboring cities without having to make a trip into town or wait for the postman to deliver papers and magazines. To the newspaper industry, radio was a dreaded new form of competition. Radio stations could transmit news to the public hours before newspapers could print it—even if several editions a day were issued. When they realized the impact radio could have on the publishing field, several large newspapers decided to get in on the broadcasting industry themselves.

The American Telephone and Telegraph Company also decided to start up a radio station. They built WEAF in a steel building in New York City, using the best equipment available at the time. But as soon as they tried to put it on the air, they ran into trouble. The steel in the building caused terrible static problems.

Before their engineers had time to figure out how to deal with that, WEAF was knocked completely off the air when a lightning strike destroyed their equipment. WEAF rebuilt their facilities—this time in a brick structure.

In the midst of all the excitement over radio, no one had devised a way to make money from broadcasting programs to the public, other than on the sale of receiving sets. But in 1922, WEAF discovered that money could be made by renting the use of their broadcasting facilities to business people—in other words, by airing commercials.

The first commercials were long and monotonous speeches, sometimes lasting for as long as 10 minutes. Their unpopularity with the public led businesses to change their advertising strategy and use shorter messages spaced between news, music, and other program features. New stations were popping up everywhere. And every one of them hoped to win the public's undivided attention. By 1927, about a thousand stations were broadcasting to the United States and Canada. This created a new problem—radio interference.

With so many stations, the airwaves were in a constant state of chaos. Anyone could operate on any frequency they chose—or even raise their power to block out a rival station's signal. To make matters worse, commercial and amateur (ham radio) stations often aired their transmissions in the same frequency range.

The government gets involved

The U.S. Congress decided to set up a Federal Radio Commission to restore some sort of order to the hopelessly overcrowded airwaves. After much debating, the commission decided that only 700 stations could reasonably operate in the available band space. Part of the solution was to require stations to lower their power at night when their signals covered a wider area.

Stations in the United States were issued call signs starting with K or W, while Canadian stations were given call letters starting with C. In the United States, the Mississippi River was eventually established as the dividing line, with K issued to stations to the west and W to the east. But KDKA, KYW, and a few other pioneers of broadcasting were allowed to keep their original station letters.

While the Federal Radio Commission was busy assigning frequencies and station letters, radio engineers did their part to help solve the problem by designing better receiving equipment that made it possible for stations to be placed closer together on the dial without causing interference to each other.

Shortwave radio develops

Ham radio operators around the world had been using shortwave frequencies for decades before the international bands became populated by commercial, religious, and government-run stations. Radio hobbyists had always taken a keen interest in these "skip" bands that allowed them to communicate with people on the other side of the earth.

Beginning in the mid-1920s, a number of commercial AM stations (including KDKA) operated shortwave transmitters to get their programs out to a wider audience. But it took World War II, and the unending quest for the latest news from the battlefields, to give shortwave listening the attention it needed to attract the general public's attention (Fig. 1-7).

1-7 Shortwave listening was popular during World War II.

With a shortwave radio, you could tune in to the Voice of America, The British Broadcasting Corporation, or other high-powered stations—and bring the most up-to-the-minute news available right into your living room. The Voice of America (Fig. 1-8), The British Broadcasting Corporation, and Radio Moscow have always been major players in the world of international shortwave broadcasting.

New types of radio shows

The National Broadcasting Company (NBC) was America's first radio network. It started with 24 stations in 1926. Several months later, CBS, the Columbia Broadcast-

1-8 This switchboard was used by the Voice of America for many years. VOA photo.

ing System, was founded. With bigger budgets than any single stations could finance, radio networks set out to create programming that people from all across the country would enjoy.

Local affiliate stations supplied the regional features such as area news, weather, and high school ball games, while networks took on the job of producing and relaying national news, sports, and entertainment shows.

The stars we all loved

Early network radio comedies, mysteries, westerns, and dramas, now grouped together as "old-time radio," gathered families around the huge wooden radio consoles each evening to listen. And not surprisingly, radio became the proving ground for many of America's best-loved entertainers. Bob Hope, Jack Benny, Red Skelton, and many others got their start on network radio shows. These were the heyday of radio.

TV changes the industry

As TV viewing began to dominate American homes, radio gradually took second place in our lives. Live music and entertainment programs were gradually replaced with recorded music, as actors, comedians, and singers migrated to the more lucrative TV and movie industries. Some thought radio was on its way out, but they were wrong.

About that time, radio became portable. Once transistors were invented, radios could be manufactured that fit into the palm of your hand. Televisions, with all their attractions, were nearly impossible to use away from home. But you could take a transistor radio with you anywhere you wanted to go. The public had discovered portable entertainment. Before long, this new product was priced so low that nearly everyone could afford to own one, and the new pocket-sized transistor radio took the nation by storm.

More program changes

Station owners were constantly looking for new ways to attract more listeners. One such strategy was audience participation programs. With the right host, you could make ratings soar.

Talk shows became popular because they gave you, the average citizen, access to the airwaves. Talk shows provided the opportunity to be heard by the community and voice your opinions on important issues of the day. Music request shows, where you could call in and have a favorite song dedicated to your special someone, were another audience grabber.

Come meet the DJ

"Remotes" were nothing but live broadcasts from an advertiser's store. But they drew in flocks of people wanting to meet their favorite disc jockey and perhaps get his autograph. And while they were there, they just might see something they'd like to buy. Stores liked the idea—and so did the listeners.

AM radio targets its listeners

When FM stereo took over the rock and country music market in the seventies and eighties, AM radio had trouble competing for listeners and advertising dollars. A new approach was definitely needed. Instead of trying to please people of all ages, interests, and backgrounds, as they always had in the past, many AM broadcasters decided to tailor their programming to a select audience and then find advertisers whose products would be likely to appeal to the group they were targeting. This selective approach led to a growing number of special-interest AM talk shows on subjects ranging from gardening and car repair to health care and money management.

FM radio changed, too. Music formats were, and are, designed to attract people of different age groups and interests, such as soft rock for adults, hard rock for the younger generation, and mixtures of old and new music designed to "please everyone in the office."

College, university, and public radio stations broadcast an endless variety of music, news, and features that would be impossible to find on commercial stations. Even children have their own programs (Fig. 1-9).

1-9 "We Like Kids" is heard on over 50 Public Radio stations.

An ever-changing medium

The early 1990s brought the Gulf War and the breakdown of communism in Eastern Europe—and made shortwave radio more popular than ever. You could buy a portable shortwave radio that fit into the palm of your hand, lift up the built-in antenna, and hear history being made. You could listen to people celebrating joyously in the streets as one repressive communist regime after another fell from power. Right before your ears, the Soviet Union ceased to exist. Almost overnight, shortwave radio stations that were once nothing but international mouthpieces for communist propaganda were transformed into lively voices of freedom, progress, and democracy.

But the struggle is far from over. There is still much work to be done. You can hear these brave people solve the many problems that confront them as their government and economic systems change from one form to another. They are carving a new, better place for themselves in the global system—and with a shortwave radio, you have a front row seat for all the action. Shortwave listening has *never* been more exciting!

No one knows where radio will go in the future.

But keep listening . . . and you'll find out.

2
Long-distance AM listening

Radio waves from space are dropping into your town tonight. They come from cities and towns hundreds of miles away. You already have all the equipment you need to receive them. All you have to do is wait until the sun goes down, tune across the band, and you'll hear them.

No, it's not science fiction, even though it might sound that way. In fact, it's been going on since the early part of the century. AM band radio signals rise to heights of over 150 miles above us, run into a layer of electrically charged gas, change directions, and speed back to earth. They touch down hundreds of miles away from the original transmission site.

And the best part is, you don't have to spend a fortune on satellite dishes or other specialized electronic equipment to hear them. The only piece of equipment you'll need to get started in this great hobby is an ordinary AM radio.

Bringing out-of-town stations in

Tuning in out-of-town stations on the AM band is an exciting and fun way to find out what's going on across the country. By simply raising the volume and S-L-O-W-L-Y turning the dial, you can hear dozens of stations as the sky wave carries their message up into outer space, back down to earth, and into your receiver. Even cheap portable radios can bring in a surprising number of stations. It's the easiest, fastest, and most economical way to get into long-distance radio listening.

AM radio has a lot to offer, no matter what your interests are. Dozens of local and national talk shows (many have toll-free phone numbers) populate the airwaves. They discuss, debate, and give vent to nearly every conceivable idea, view, and opinion. You can hear, and (if you're lucky) get on the air and talk to celebrity guests—famous writers, actors, sports heroes, politicians, and people making the news in general.

Want more? There are plenty of action-packed ball games almost every night of the week. You'll hear music that ranges from rock and country to classical, gospel, ethnic, and oldies. Religious broadcasters of every persuasion preach their message to anyone that might be listening. And coast-to-coast truckers' shows, complete with country music requests, road information, and weather forecasts, keep drivers company as they haul their loads down the highway all night long.

When the sun goes down tonight, tune slowly across the AM band, and you'll see how much you've been missing. Turn the volume up a few notches and explore the "empty" spaces between your local area broadcasters.

Stations that you never knew existed will assert their voices through your radio speaker, sometimes almost as clearly as the hometown stations you've been tuning in for years. Nearly every metropolitan area is home to at least one 50,000-watt station, and if you live within a thousand miles of it, you can probably hear it. Once you get the hang of tuning in out-of-town stations (and it won't take long), you'll have dozens of new information and entertainment options.

How AM signals travel

When radio waves leave the antenna, they rush away from the transmission tower at the speed of light (Fig. 2-1). Station engineers try their best to keep as much of the signal as possible down close to the earth, where the listeners are.

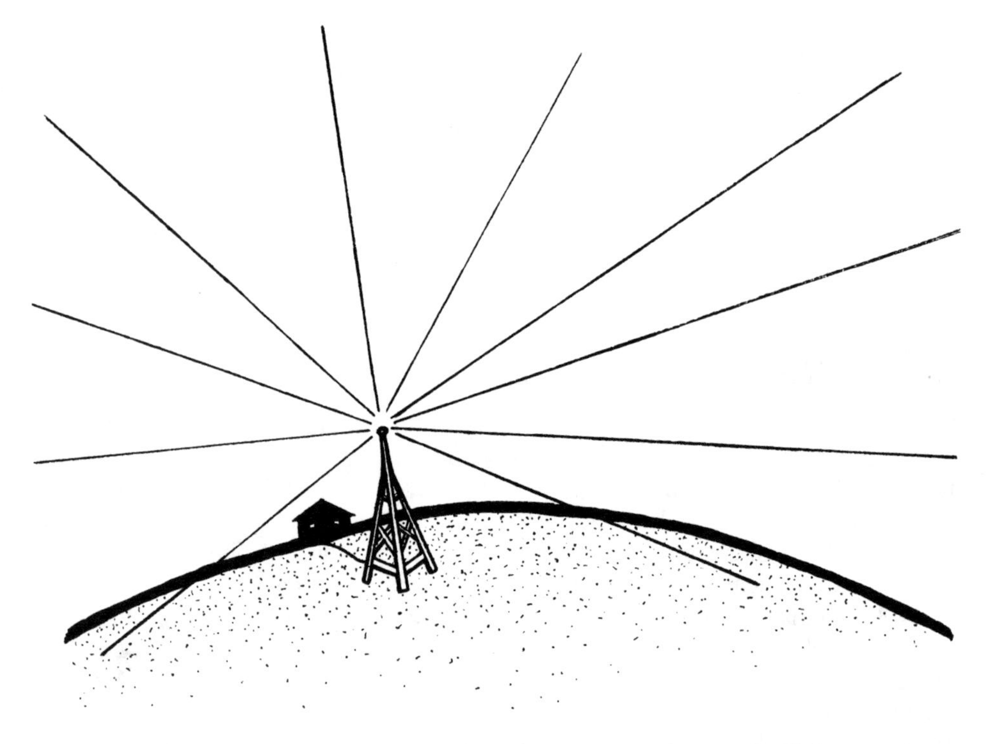

2-1 Radio waves travel away from the antenna at the speed of light.

But no matter how well the antenna system is designed, some radio waves do go astray and shoot up into the sky. In the daytime, these wayward signals don't get very far. They are absorbed as they pass through the D-layer of the ionosphere, a highly charged layer of gas about 40 to 50 miles above us. (During daylight hours, the D-layer is continually being charged by the sun's radiation.) All the electricity used to generate daytime sky waves is wasted, as they are drained of all their energy while moving through the D-layer and are never heard from again (Fig. 2-2).

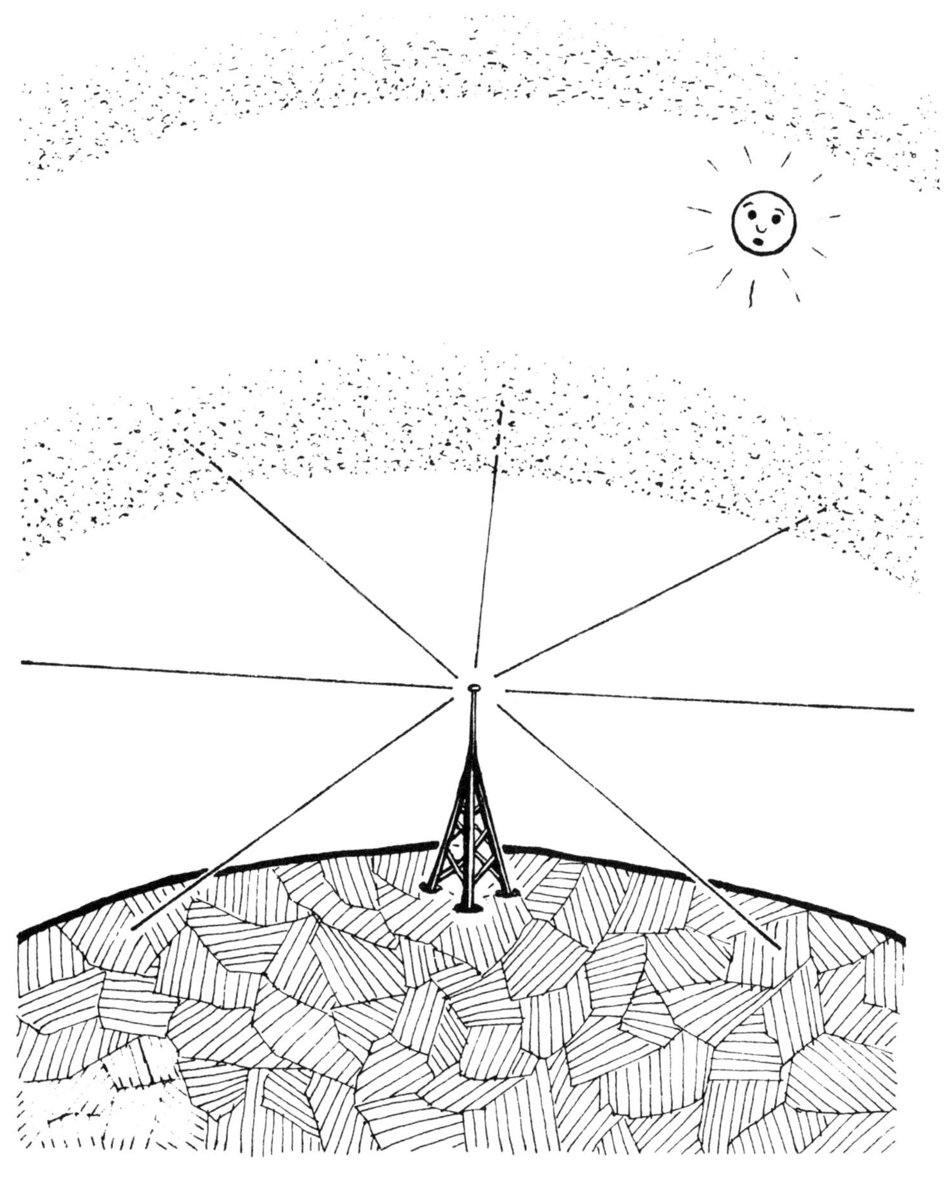

2-2 During the daytime, the D-layer of our ionosphere absorbs AM sky waves.

But at night, everything changes. Once the sun goes down, AM radio waves are free to travel. When radio waves get to an altitude of about 150 miles above us, they reach another layer of electrically charged gas known as the *F-layer* of the ionosphere, which has a completely different effect on them. Molecules are farther apart that high in space, and instead of destroying radio waves, the F-layer bends them and directs them back down to earth (Fig. 2-3) to the delight of long-distance radio listeners everywhere!

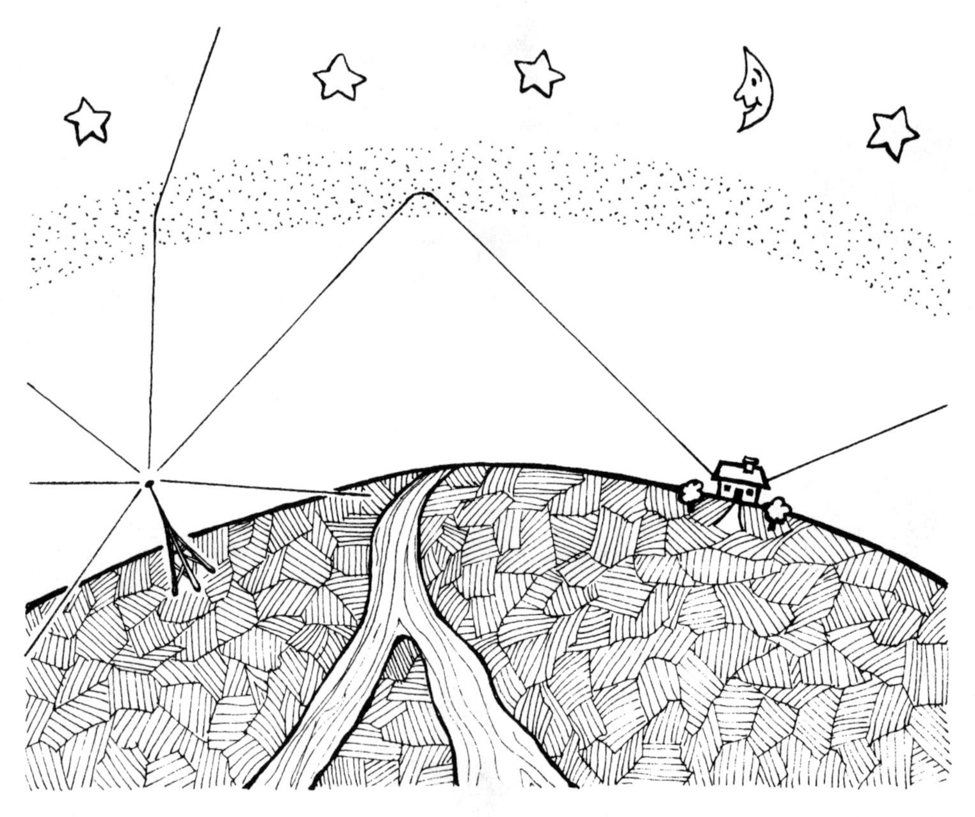

2-3 At night, AM waves can "skip" hundreds of miles.

Clear channel stations

Local and regional broadcasters want as much of their signal strength as possible to go toward providing listeners in the immediate area with music, news, and talk shows. But clear channel stations have a slightly different goal. They not only want to get a good signal out to their local audience, they want to reach as many miles as they possibly can with their powerful 50,000-watt signal. So when evening comes, they adjust their antennas to allow more of the signal to go up into space and hit the ionosphere at the right angle to bring clear channel programs to out-of-town listeners, hundreds or possibly thousands of miles away.

2-4 David Brudnoy, of WBZ 1030 in Boston, Massachusetts.

Clear channel nighttime talk show hosts, such as David Brudnoy (Fig. 2-4), on Radio WBZ 1030 in Boston, Massachusetts, receive most of their calls from out-of-town listeners. WBZ is heard in more than 35 states and in several Canadian provinces. Table 2-1 gives the station letters and frequencies of every 50,000-watt station in the United States and Canada.

Table 2-1. 50,000-watt AM stations in the United States and Canada.

U. S. stations

Station	kHz	City and state/province
WGTO	540	Pine Hill, FL
KFI	640	Los Angeles, CA
KYAK	650	Anchorage, AK
WSM	650	Nashville, TN
KTNN	660	Willow Rock, AZ
WFAN	660	New York, NY
KOBI	670	Boise, ID
WMAQ	670	Chicago, IL
KNRB	680	San Francisco, CA
WRKO	680	Boston, MA
WPTF	680	Raleigh, NC
KUET	700	Glendale, AZ
WLW	700	Cincinnati, OH
KIRO	710	Seattle, WA
WAQI	710	Miami, FL
WOR	710	New York, NY
KDWN	720	Las Vegas, NV
WGN	720	Chicago, IL
KTRH	740	Houston, TX
KCBS	740	San Francisco, CA
WWNZ	740	Orlando, FL
KFQD	750	Anchorage, AK
WSB	750	Atlanta, GA
KFMB	760	San Diego, CA
WJR	760	Detroit, MI
KKOB	770	Albuquerque, NM
WABC	770	New York, NY
KROW	780	Reno, NV
WBBM	780	Chicago, IL
KGO	810	San Francisco, CA
WGY	810	Schenectady, NY
KCBF	820	Fairbanks, AK
WBAP	820	Ft. Worth, TX
WCCO	830	Minneapolis, MN
WHAS	840	Louisville, KY
KOA	850	Denver, CO
WHDH	850	Boston, MA
WWL	870	New Orleans, LA

Table 2-1. Continued.

Station	kHz	City and state/province
WCBS	880	New York, NY
WLS	890	Chicago, IL
KFRE	940	Fresno, CA
WINZ	940	Miami, FL
KOMO	1000	Seattle, WA
WLUP	1000	Chicago, IL
WINS	1010	New York, NY
KCKN	1020	Roswell, NM
KDKA	1020	Pittsburgh, PA
KTNQ	1020	Los Angeles, CA
KTOW	1030	Casper, WY
WBZ	1030	Boston, MA
WHO	1040	Des Moines, IA
KYW	1050	Philadelphia, PA
WEVD	1050	New York, NY
KXN	1070	Los Angeles, CA
KRLD	1080	Dallas, TX
WTIC	1080	Hartford, CT
KAAY	1090	Little Rock, AR
KING	1090	Seattle, WA
WBAL	1090	Baltimore, MD
KFAX	1100	San Francisco, CA
WWWE	1100	Cleveland, OH
KFAB	1110	Omaha, NE
WBT	1110	Charlotte, NC
KMOX	1120	St. Louis, MO
KPNW	1120	Eugene, OR
KWKH	1130	Shreveport, LA
WNEW	1130	New York, NY
KRAK	1140	Sacremento, CA
WRVA	1140	Richmond, VA
KSL	1160	Salt Lake City, UT
KVOO	1170	Tulsa, OK
WWVA	1170	Wheeling, WV
WHAM	1180	Rochester, NY
KEX	1190	Portland, OR
WOWO	1190	Ft. Wayne, IN
WOAI	1200	San Antonio, TX
WOGL	1210	Philadelphia, PA
WKNR	1220	Cleveland, OH
KSTP	1500	St. Paul, MN
WTOP	1500	Washington, DC
KGA	1510	Spokane, WA
WLAC	1510	Nashville, TN
WSSH	1510	Boston, MA
KOMA	1520	Oklahoma City, OK

Table 2-1. Continued.

Station	kHz	City and state/province
WWKB	1520	Buffalo, NY
KFBK	1530	Sacremento, CA
WCKY	1530	Cincinnati, OH
KXEL	1540	Waterloo, IA
WPTR	1540	Albany, NY
WQXR	1560	New York, NY
KBLA	1580	Los Angeles, CA
KCWW	1580	Tempe, AZ
	Canadian stations	
KRVN	880	Lexington, NB
CBK	540	Regina, SK
CFNB	550	Fredericton, NB
CKY	580	Winnipeg, MB
CKYC	590	Toronto, ON
CHED	630	Edmonton, AB
CBF (French)	690	Montreal, PQ
CBU	690	Vancouver, BC
CBL	740	Toronto, ON
CBX	740	Edmonton, AB
CHQR	770	Calgary, AB
CFCW	790	Edmonton, AB
CIGM	790	Sudbury, ON
CHRC (French)	800	Quebec, PQ
CKLW	800	Windsor, ON
CJBC (French)	860	Toronto, ON
CHQT	880	Edmonton, AB
CHML	900	Hamilton, ON
CHDQ	910	Drumheller, AB
CFBC	930	St. John, NB
CJCA	930	Edmonton, AB
CJYQ	930	Harbor Glace, NF
CBM	940	Montreal, PQ
CFAC	960	Calgary, AB
CKNW	980	New Westminster, BC
CBY	990	Corner Brook, NF
CFRB	1010	Toronto, ON
CBR	1010	Calgary, AB
CHUM	1050	Toronto, ON
CBA	1070	Moncton, NB
CFCN	1070	Calgary, AB
CKWZ	1130	Vancouver, BC
CKOZ	1150	Hamilton, ON
CKDA	1220	Victoria, BC
CFRN	1260	Edmonton, AB
CJMS (French)	1280	Montreal, PQ
CKCV (French)	1280	Quebec, PQ

Table 2-1. Continued.

Station	kHz	City and state/province
CIWW	1310	Ottawa, ON
CHQM	1320	Vancouver, BC
CFUN	1410	Vancouver, BC
CJCL	1430	Toronto, ON
CJVB	1470	Vancouver, BC
CKLM (French)	1570	Montreal, PQ
CBJ (French)	1580	Chicoutimi, PQ

Small town radio

Small town stations are more likely to be locally owned than high-powered outlets and the announcers are usually neighborhood residents getting their start in broadcasting at the hometown station they grew up with But don't let this make you think that small town radio is boring! In its own unique way, a small town station can be just as interesting as any station on the dial.

For one thing, formats are more informal and relaxed. Because a small town station is often the only broadcaster in town, they do their best to serve the entire community. On the other hand, big city stations narrowcast to win a select audience—such as the youth market, the financial community, or one of the area's many ethnic groups. Small town radio tries to appeal to people of all ages and interests. They are the station that the teenager, the grocery store owner, and the farmer tune to for entertainment, news, weather forecasts, and to find out what's going on in the community in general.

As far as music goes, you're likely to hear country or middle-of-the-road pop songs. Talk shows are for discussing local issues, exchanging recipes and gardening tips, or running swap shops on the airwaves. You'll also hear plenty of announcements—church socials, farm prices, school lunch menus, lost pets, and, of course, birthdays and anniversaries of area residents.

If you slowly scan the AM band, you're likely to find a number of low-power stations that you've never noticed before. Even in the daytime before the skip waves come in, you can probably hear at least a handful if you turn up the volume and check between local stations. While you can find low-powered stations all across the AM band, the majority operate between 1230 and 1490 kHz in the United States.

As you listen, you'll discover that every small town station has a personality of its own, a reflection of the community it serves. Find one you enjoy, and let them hear from you. They'll be happy to know that they have you for a listener.

Hello . . . you're on the air

For many long-distance AM listeners, talk shows are what nighttime listening is all about. The great hosts, interesting guests, and opinions of your fellow citizens have made talk shows one of AM radio's most popular attractions. And, if you pick up the telephone and punch in the number, it could be *you* voicing your opinion over the airwaves for the listening public to hear.

Even though the phone lines are open to everyone, it has been estimated that only 1 or 2 percent of talk show listeners ever call in. They just sit back and miss out on their opportunity to speak with the guest, air an opinion, confront a host they disagree with, or bring up an entirely new topic—and be heard by an audience of untold thousands.

Some talk shows are general interest, meaning that almost any topic can come up for discussion. But there are also a good number of special interest shows on the air, hosted by experts in fields such as car repair, home fix-ups, gardening, pet care, family problems, sports, and so forth. If you have a question or need some help to solve a problem, the answer is only a phone call away.

National network talk shows use modern satellite technology to deliver their programs to hundreds of stations throughout the United States. With an audience that numbers into the millions, an endless supply of famous authors, celebrities, scientists, politicians, sports stars, and other interesting people are always available to the host. But getting through the busy phone lines can be a real challenge. So let's look at some easier targets:

Fifty-thousand-watt clear channel stations have enough power to get their signal out several hundred miles—roughly half of the country can hear them after dark. They, too, have a good supply of interesting guests. Most have toll-free phone numbers, and getting through to them can take some patience, but you can usually get on the air if you keep trying long enough.

Let your voice be heard

If you've never been on a talk show before, local programs are an even better place to start. Getting in is fairly easy, especially if you live in a small to medium-sized community. Local talk show hosts are always happy to hear new voices, because one of the biggest complaints they get is that the same people call in day after day.

When you call a talk show, you'll probably run into a screener, a person the station employs to answer the phone for the host and find out who's calling and what they want to talk about. Generally, they ask your first name, location, and what you would like to discuss. Then, you are put on hold to wait for your turn to go on the air. Screening helps the host to organize his program and cuts down on prank callers.

Many talk shows use a five-second delay, so it's a good idea to switch your radio off when it's your turn to talk. It can be very confusing and distracting for you, as well as the host, to hear an instant replay of everything that's just been said.

Most people have a little stage fright the first time they go on the air, but don't let that stop you! After you've done it a few times, getting on a talk show to express your opinion becomes as easy and natural as picking up the phone to call a friend.

Twilight listening

The majority of stations on the AM band run less than 5,000 watts. Of course, you can hear them if you live in or anywhere near the area they're supposed to cover. With a good receiver and the right skip conditions, you can often hear them well beyond their normal broadcasting range after dark.

The Federal Communications Commission requires many of these low- to medium-power stations to reduce their signal strength at night to avoid interfering with other broadcasters. So every time the sun rises and sets, you are presented with a unique listening opportunity! At twilight, you can hear a good number of stations when they are running their full legal daytime power, and enough of the ionosphere's D-layer is absent to let evening skip conditions come into play. While twilight skip is never as strong as nighttime skip, it might be your only chance to hear many low-power AM stations. In the evening, stations to the east will sign off first, with stations to the west following later. And in the morning, stations to the east are the first to return to the airwaves as the sun rises over the horizon.

Twilight is also a good time to try for late afternoon/early evening "rush hour drive-time" shows on Clear Channel stations. If you live on the East Coast, for example, you might want to turn up the volume and see if you can tune in Les Jameson (Fig. 2-5), the late afternoon British talk show host on WLAC in Nashville, Tennessee. Les Jameson has a popular show and a very interesting background. He was a stuntman/circus performer in his native country of England. Later, in the United States, he became an actor, director, producer, public speaker, and of course, a radio talk show host! He's been on Radio WLAC 1510 AM for more than 12 years.

Noise and static—what you can and can't do about it

When you listen to local stations, noise and interference are seldom, if ever, a problem. But when you're trying to hear out-of-town stations, the signals you pick up are weak from their long journey. So local noise and interference can easily work against you, blocking out the very signals you want to hear. Figure 2-6 shows some common causes of radio noise.

Thunderstorms

The most common (especially during the warmer months) and most powerful cause of radio noises is a thunderstorm. While you might barely notice static crashes on a local broadcast, they can easily ruin reception of out-of-town stations.

Fluorescent lights

Fluorescent lights are another big problem for long-distance listeners. If you—or sometimes even one of your neighbors—has one, you'll probably hear a buzzing sound on your radio whenever it is in use. Here are some solutions:

- Remove the light.
- Take your radio to another part of the house.
- If you find out it's coming from a neighbor's house, explain the problem and offer to replace the fluorescent light with another type of fixture.

2-5 Les Jameson can be heard on WLAC 1510, Nashville, late afternoons.

Tools, appliances, etc.

Power tools and motor-driven appliances are another major source of radio noise. Power saws, routers, electric razors, and kitchen blenders are among the offenders.

2-6 Many things can cause noise on your radio.

Because these types of appliances are usually operated for only short periods of time, the best thing to do is wait it out or take your radio to another part of the house where the buzz isn't as loud.

Power lines

Sometimes problems with power lines can cause a constant buzz on AM (and shortwave) radios. If you're hearing a buzz on your radio 24 hours a day, contact your power company, explain the problem, and ask them to check around the neighborhood for the source of the noise.

Etc. . . . etc. . . . etc.

- No matter what type of music you like, you can probably find an AM station that plays it. Of course, there's always rock and country music. But if you look around the band, you can also find stations that play oldies, gospel, jazz, classical, bluegrass, or ethnic music.
- If you like to call in to out-of-town talk shows, it's a good idea to keep a list of telephone numbers of the programs you regularly participate in. It saves lots of time and frustration when you have an opinion to express but the host never seems to get around to broadcasting the number.
- Sports fans can always find plenty of action on the AM band. The airwaves are filled with the excitement of high school, college, and professional teams.

And if that's not enough, sports-oriented talk shows are constantly interviewing popular players, managers, and coaches. Write to your favorite team's office for a schedule and list of stations that carry their games. Enclose a self-addressed, stamped envelope for a quick reply.

- Want to know what radio was like in its early years? Find a station that plays tapes of those great comedies, dramas, westerns, and mysteries that first hit the airwaves in the thirties, forties, and fifties. They're just as entertaining as they were decades ago.
- Religious programs are aired seven days a week on an increasingly large number of AM stations. Everything from traditional church services to more unorthodox approaches to the spiritual life get their share of airtime, as do inspirational music programs and Christian call-in shows.
- Taking a trip? If you're driving late at night, why not tune in one of the 50,000-watt truckers' stations. You'll hear weather and road reports for every major highway across the country, and plenty of country music to go along with it!
- If you're traveling to a city within several hundred miles of your home, you can probably hear them on the radio before you go. You'll find out what's going on, and maybe even discover some special events you'll want to attend when you arrive.
- Some long-distance AM listeners enjoy tracking weather systems across the country by radio. You can not only keep up-to-date on the weather, but also trace the effect it is having on communities as it travels from city to city. You'll hear reports of icy roads, fallen power lines, flooded streets, airport delays, school closings, and so on.
- *The World Radio TV Handbook*, published by Billboard Publications, Inc., lists all Canadian stations running at least 100 watts and all United States stations running at least 5,000 watts. It also lists station addresses and gives information on the type of programming they broadcast. (The WRTH contains information on many foreign stations as well.)
- If you'd like to buy a book listing every AM station in the United States and Canada, you can get it from the National Radio Club! Every year, they publish a new, revised edition of the *NRC AM Radio Logbook*. This book is more than just a station list. It's filled with up-to-date information on how much power stations run, what type of programming they carry, and so on.

  The National Radio Club publishes a monthly club bulletin, *DX News* (Fig. 2-7), and arranges for AM stations across the country to run special late night test transmissions, making it easier for their members to hear them. For more information and a sample copy of *DX News*, write to:

  National Radio Club, Inc.
  P.O. Box 5711
  Dept. ALM
  Topeka, KS 66605-0711
  (please enclose a first-class stamp).

DX News

•Serving DX'ers since 1933•

Volume 59, No. 21 - March 2, 1992 (ISSN 0737-1659)

Inside...

CPC Test Calendar

Call	Freq	Date	ELT Time
KBBS	1450	Mar 1, 1992	0200-0400
KXOL	1320	Mar 2, 1992	0130-0200
WRIX	1020	Mar 2, 1992	0200-0300
KALT	900	Mar 7, 1992	0600-0630
KPCR	1530	Mar 8, 1992	0100-0600
WZAO	1370	Mar 8, 1992	0530-0600
KMTI	650	Mar 9, 1992	0230-0300
KTNS	1090	Mar 9, 1992	0300-0400
WKYB	1000	Mar 16, 1992	0200-0300
WCIL	1020	Mar 16, 1992	0300-0330

From the publisher ...

Don't forget that all DDXD loggings go to Bill Hale until we select additional editors. Jerry Starr's AM Switch should return next week.

I think you'll enjoy the centerfold this week, and thanks to Ray Nemec for providing the copies of the 1922 maps. Note the March 25, 1924 verie (?) from PWX, Havana ... does anyone have an older verie in their possession? If so, we'd like to print it. Send a clean photocopy of it to Topeka.

The NRC is looking for one or more NRC members to represent and promote the NRC at the Dayton (OH) HamVention, April 24-26. Volunteers would deliver sample copies of DXN and other materials and help man an ANARC club table, and possible help conduct a listeners' forum April 26 from 11:15-1:00 pm. At the least, you'd deliver materials to the table. If interested, write to Topeka.

Speaking of the American Association of North American Radio Clubs ... this entity, once an umbrella organization for NA clubs, shows signs of coming back to life under the guidance of Interim Coordinator Richard D'Angelo. Your publisher is now the ANARC rep. I recently sent a long letter detailing my concerns to Richard, not the least being that the board and candidates-elect were heavily dominated by members of shortwave clubs, especially one club. The NRC has no intention of pulling out of ANARC, as did ODXA several years ago, as D'Angelo assures member clubs that they, not one person, will determine the future of ANARC. The NRC will definitely contribute to discussion concerning ANARC's future, and NRC members who have concerns and suggestions should send them to me for forwarding.

Albert S. Lobel - P. O. Box 26762 - San Diego, CA 92196-0000 will verify that you heard his business's commercial on KBNN-100.1 if you send him a cassette tape with the commercial recorded on it. Include a stamp for his letter and return postage for the cassette. Note that this is not an actual station reception report.

The Hollow State Newsletter is up and running again, with Dallas Lankford editing and Ralph

(continued on page two)

DX Time Machine

Receptions from the DX log of Max Watkins, Tecumseh, NE: **1923: Jan. 10: WOT (?), Indianapolis, IN; KSD, St. Louis, MO; WOC, Davenport, IA; Jan. 11: KFAF, Denver, CO; WHB, Kansas City, MO; WFAV, Lincoln, NE.**

From the pages of *DX News*

50 years ago ... **From the Feb. 28, 1942 *DX News:* NRC President Joe Becker reported that his BCB veries totaled 1446; Editor Ray Edge's total was 640, and Ernie Cooper's was 974 "and plugging for 1000".**

25 years ago ... **from the Mar. 4, 1967 DXN: Mort Meehan, Van Nuys, CA, noted that he joined the NRC in 1934, the same year that he joined the NNRC ... Glenn Hauser, Albuquerque, NM noted that KLOS-1580 went off the air in early Jan.**

10 years ago ... **from the Mar. 1, 1982 DXN: John Clements, Sun Valley, CA visited Louisville HQ Feb. 10 ... Ron Schatz's "Cuban MW Broadcasting in Perspective" was published in this issue.**

2-7 The cover of *DX News*, published monthly by the National Radio Club, Inc.

Collecting station souvenirs

Almost everyone involved with long-distance radio listening enjoys collecting souvenirs, bumper stickers, key chains, and other promotional items from the stations they hear (Fig. 2-8). Many AM stations (especially the larger ones) have printed cards to send out, too, as a reward to listeners that let them know how well their signal is coming in. They usually feature the station's call letters and give information on the power level, antenna system, and so forth.

2-8 AM and FM stations have great souvenirs for their listeners.

Radio hobbyists refer to these prizes as QSL cards. QSL is a ham radio term meaning "I verify reception"—or, in other words, "Yes, you heard us!" Even smaller stations that don't have a supply of printed QSL cards will usually mail out a letter asking whether you heard their broadcast, especially if you enclose a self-addressed stamped envelope.

Writing to AM stations

To get a QSL card, you'll need to make out a brief report about what you've been listening to. In the world of radio, this is known as a reception report. When the station personnel receive it, they check with their program log to make certain that you really heard them. A reception report should include:

- Your name and address (so they'll know where to mail your QSL card).
- The time, preferably in the station's own local time.

- The month and day of the year.
- The frequency (number on the dial) where you heard them.
- The strength of the signal (loudness).
- If there was any interference.
- A brief summary of the program you heard.

It also helps to tell a little about yourself, your listening equipment, and your opinion of their programming.

Let's say that you heard KMOX on 1120 kHz. It's an easy catch for most people in the United States and Canada living east of the Rocky Mountains. At the top of the hour, the announcer says, "This is KMOX. The time is 1:00 A.M. in St. Louis, Missouri." In just one sentence, they've told you three of the things you'll need to know to get a QSL card: the station letters, the location, and the local time. (See, I told you it was easy!) Now, the news comes on. Jot down the network. It will probably be followed by a commercial, public service announcement, or a station identification announcement. Note that in your report, too.

Okay, it's time to get back to the talk show. The host is Jim White, and he's having an open line program tonight. Make sure to write down his name, the program sponsors, and some of the topics brought up by his callers.

If you don't know the address of a station you want to contact, the library can probably help you. They have books listing radio and TV stations throughout the country. Out-of-town telephone books, available at some libraries, are another good source of addresses. And if you have a copy of *The World Radio-TV Handbook,* you can look up the address of most AM stations operating in the United States and Canada.

If the station you want to QSL is located in a small to midsize town, the letter will usually reach its destination even if you have only the station letters, city, and zip code (call your post office for zip code information). Some stations answer QSL requests faster than others, but you'll always increase your chances of receiving a speedy reply if you enclose a self-addressed, stamped envelope. Replies can still take anywhere from weeks to months. So be patient!

3
Shortwave radio

WOULD YOU LIKE TO EXPLORE THE WORLD, VISIT NEW PLACES, LISTEN TO EXOTIC music, find out about cultures most people didn't even know existed, and get all the vital background information you've been missing on history-making events overseas?

Your passport to the world

Unless you have a millionaire's travel budget, shortwave radio is the way to go. It's a real bargain. You can investigate any part of the world you choose—as often as you like—and you won't even have to buy one airline ticket.

Believe it or not, all the equipment you'll need to explore these fascinating places will easily fit into the palm of your hand. And prices start at under $100. By simply switching on your shortwave radio, pushing a few buttons or turning a dial, you can tune in broadcasts from around the world and find out what's going on in nearly every corner of our planet.

Imagine how much fun it would be to turn on your new world band radio and dart from Moscow to London, then down to Australia and over to Canada. With a few more turns of the knob, you can hear what's going on in New Zealand, Brazil, Germany, Taiwan, Israel, Nigeria, Costa Rica, South Africa, Canada, and almost every other nation.

And here's the best part: Because most of the stations you'll be hearing broadcast at least some of their programming in English, you don't have to learn a foreign language to understand them.

> Having a shortwave radio today is crucial. It's the easiest, fastest, and cheapest way to get a worldwide perspective of what is happening on the planet we live on.

> No one news service can tell it all. With shortwave radio, you have dozens of alternative sources of information. And the more we know about what's happening around us, the better it is for everyone.
>
> Al Weiner, Radio Newyork International

Millions of people around the world own shortwave radios. They can tune in stations from neighboring countries one minute, and the next minute they'll hear signals that have traveled halfway around the planet to reach them. And once you own a shortwave radio, you can too!

What's on the international bands?

What's actually on international bands? In a word— *everything*! You'll hear national and world news, interviews and opinions, music, press reviews, and mail bag programs—where letters from listeners are read to a worldwide audience. There are government-run stations, religious broadcasters, ham radio operators, ship-to-shore communications, weather forecasts, and privately owned commercial shortwave stations that operate in much the same way as AM and FM stations in the United States and Canada.

Every major country on earth, as well as many of the smaller ones, has at least one international shortwave radio station. If you listen on a regular basis, you will gain an insight into the culture, problems, and ideals of a nation that you just can't get through other media outlets.

The evening network newscasters have only a half hour to tell you what's going on in your country, and fill you in on the rest of the world as well. They decide what stories you are most interested in and let all other international news stories go unreported. Newspapers, even in big metropolitan papers, only give you a paragraph or two of information on events that can change the lives of millions of people overseas.

But when you own a shortwave radio, it's all up to you. You are, in a sense, the program director. You decide what part of the world you want to know about, you tune that area in, and then you are the audience that sits back and listens.

And if you like, you can be the media critic that contacts the station to tell them what you do or don't like about their broadcast. You can even ask a question or suggest a topic for a future program—and be much more likely to have your letter taken seriously than if you directed it to one of the major news networks here in North America.

But there's much more to shortwave listening than news: There are music shows that ask for (and play) your requests, business reports that let you know about the latest trends and developments long before they hit the American press, contests (with T-shirts, stamps, books, station souvenirs, and so forth for prizes) that you really have a chance at winning, tourist information on places you might like to visit, features about life overseas, religious programs of every type imaginable, news of international developments in science and technology, and rugby and soccer games that move so fast the announcer can barely keep up with them. Shortwave stations try for a balance of all the elements that attract a worldwide audience. They are constantly working to get the right combination of news, information, features, and music to attract people like you, and keep you tuning in day after day.

The radio you'll need

Of course, if you want to hear shortwave stations, you'll need a shortwave radio! If you have a multiband "police radio" around the house, see if it has bands marked SW1, SW2, and so forth. While these radios don't bring in stations nearly as well as radios primarily designed for shortwave reception, they can certainly give you a taste of international listening.

When you decide to buy a *real* shortwave radio, you have several things to consider. First, how much money can you afford to put into your listening activities? Several nice portables are available for less than $130, and some sets are even on sale for under $50. The Grundig Yacht Boy 206 (Fig. 3-1) is a nice starter radio.

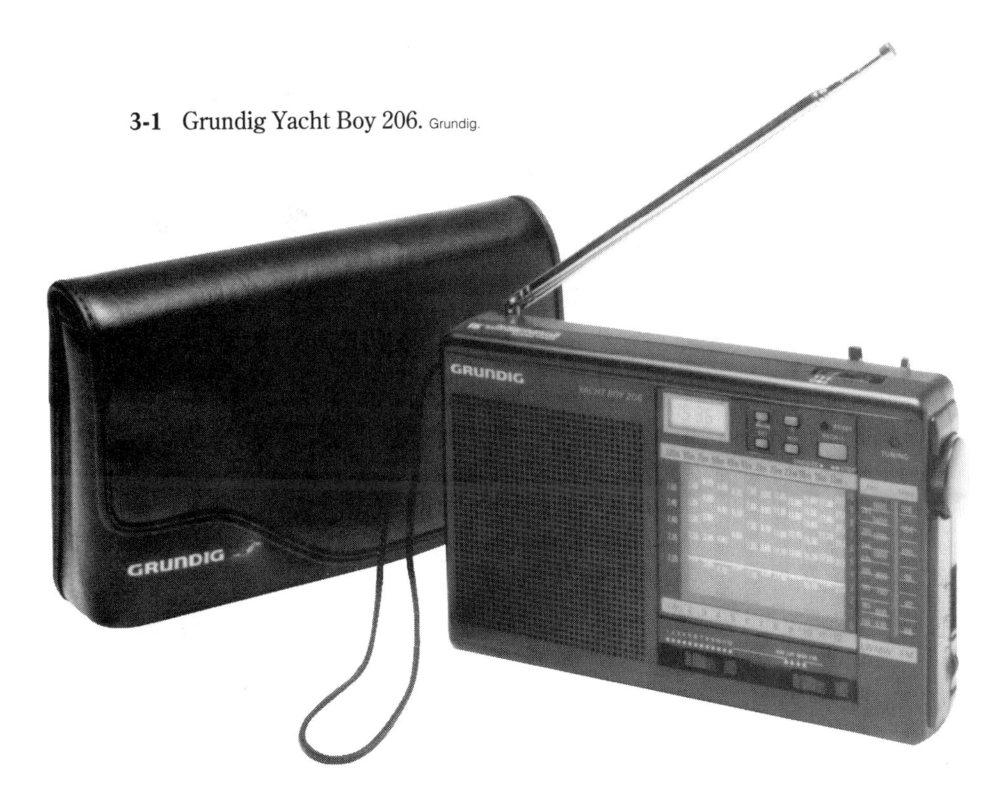

3-1 Grundig Yacht Boy 206. Grundig.

If you're willing to put a little more money into your international listening adventure, you can buy a radio with digital readout, making it much easier to tell exactly which frequency you are on. Many portable radios in the medium price range also have memories, so you can punch in the frequencies of your favorite stations and find the programs you want in a matter of seconds. Figure 3-2 is a Realistic DX 390 World Band Radio, available at Radio Shack stores nationwide. It has digital readout, a 45-station memory, and weighs less than five pounds with batteries. And finally, if you have several hundred dollars or more to invest, you might want to think about buying a tabletop communications receiver, such as the Icom IC-R72, shown in Fig. 3-3.

3-2 Realistic DX 390. Radio Shack.

3-3 Icom 1C-R72 Communications Receiver. Icom photo.

Communications receivers come with many advanced features portables don't have. While all those exciting high-tech controls are an advantage to seasoned operators, these features can frustrate the beginner, because they make the set more complicated to operate. Also, communications receivers usually require external antennas (all portable shortwave receivers have their own built-in, extendable antennas).

If you're new to shortwave, it's better to start with a low- to medium-priced portable, then buy a communications receiver when you've had some experience. Chapter 5, "And now for the details," tells more about the features to look for when you make your first shortwave radio purchase.

A world of exciting programs

- If you're new to shortwave listening, you'll be surprised to discover that some stations don't sound very "foreign" at all. Quite a few countries are currently trying to Americanize their programs to attract listeners in the United States and Canada. Some have even gone as far as importing announcers from North America to host their programs.
- You can also hear a good number of British announcers working for shortwave stations throughout the world.
- In many countries, shortwave radio is just as common as AM and FM are to us. If you own a good receiver, you can hear domestic service shortwave programs from neighboring countries; and when conditions are right, you can hear domestic broadcasts from nations thousands of miles away.
- Almost everyone is having to deal with budget cutbacks these days. And shortwave stations are no exception. New Zealand, Canada, and a few other countries are currently airing some of their domestic service programs over their international service transmitters. This helps them to save on the costs of producing separate programs for overseas listeners. And it gives you a chance to gain a unique insight into the lives of the people in the part of the world you're monitoring.
- Because shortwave stations have only so much time to broadcast all the language services they carry, they do their best to beam programs in your language to your part of the world at the time you're most likely to be listening. If you live in North America, you'll find that most stations air their English broadcasts during the evening hours, although you can find some programs in English any time of the day you want to listen.
- Dozens of South American and African domestic service programs can be widely heard throughout the world, especially at night. Because they are intended for a local audience, nearly all are in the native language of the country. But it's fun to listen to the rhythmic music, even if you can't understand the words.

  Most North American listeners can hear at least one of these domestic Canadian stations on the 6 MHz band, relaying their AM programming over shortwave.

Shortwave	Frequency (MHz)	Relays AM Station
CFCX	6.005	CFCF Montreal
CFVP	6.030	CFCN Calgary
CFRX	6.070	CFRB Toronto
CKFX	6.080	CKWX Vancouver
CHNX	6.130	CHNS Halifax

Figure 3-4 is a souvenir card from CFRB/CFRX.

3-4 CFRB/CFRX card.

- Nearly all shortwave radios have AM and FM bands. And because shortwave radios are better designed than those intended for only local reception, you'll hear dozens of out-of-town stations on the AM band that just won't come in on your $10 radio. And when an FM skip comes in (see chapter 8), you'll have a better chance at hearing it.
- Business people around the world know the many advantages of owning a shortwave radio. You get financial reports from every corner of the globe. You hear about the latest trends in business and marketing. And when you take a trip overseas, your shortwave radio can go along in your briefcase and keep you up-to-date.
- Did you ever wonder what people in foreign countries think about events happening here in North America? Buy a shortwave radio, and you'll surely find out.
- The "Voice of America" and "Radio Canada International" keep the rest of the world informed about the daily goings-on in North America.

- In addition to what you hear about us from our own shortwave outlets, important happenings in the United States and Canada are nearly always included in most overseas stations' daily press reviews and editorials.
- If you'd like to find out more about the country of your ancestors, shortwave radio is a great place to start. You'll hear news, features, entertainment, national music, cultural programs, and travel tips.
- Students often discover that shortwave listening is a big help with school projects. Language, geography, world culture, and history assignments come to life when you have a shortwave radio around.
- If you're trying to learn a foreign language, shortwave is a great way to hear native speakers on a daily basis. The following stations offer on-the-air language courses and are happy to mail out lesson materials free of charge. All you have to do is write in (using the addresses in Appendix A) and request them.

Radio Beijing	"Learn to Speak Chinese"
Deutsche Welle	"German by Radio"
Radio Finland	"Starting Finnish"
Voice of Free China	"Let's Learn Chinese"
Radio Japan	"Let's Learn Japanese"
Radio Moscow	"Russian by Radio"
Spanish Foreign Radio	"Spanish Course by Radio"

- Going on a trip overseas? Shortwave radio is a great way to find out more about the country you plan to visit. You'll get the most up-to-date information available on national and regional events, weather, and activities. You might even find out about an intriguing new tourist spot you'll want to add to your agenda.

Shortwave signals skip around the world

While AM and FM stations have a mostly local audience, high-power shortwave stations broadcast to listeners around the globe. To make sure the signal reaches their intended audience, they have to plan carefully which frequencies to use.

Only at night can the AM band be used for long-distance broadcasting. But most shortwave frequency bands can be used during daylight hours. Because of their shorter wavelength, signals can pass directly through the D-layer without losing too much energy, then reach the E-, F1- and F2-layers of the ionosphere (Fig. 3-5), and "bounce" back to earth hundreds of miles away.

When radio waves leave the antenna, they hit the ionosphere at many different angles. Some waves skip off the E-layer, while others go as high as the F-layer before returning to earth. Multihop skips are also common in the shortwave frequency bands (Fig. 3-6).

Where to listen

The International Telecommunications Union has set aside several portions of the shortwave radio frequency spectrum for international broadcasting. Other parts of

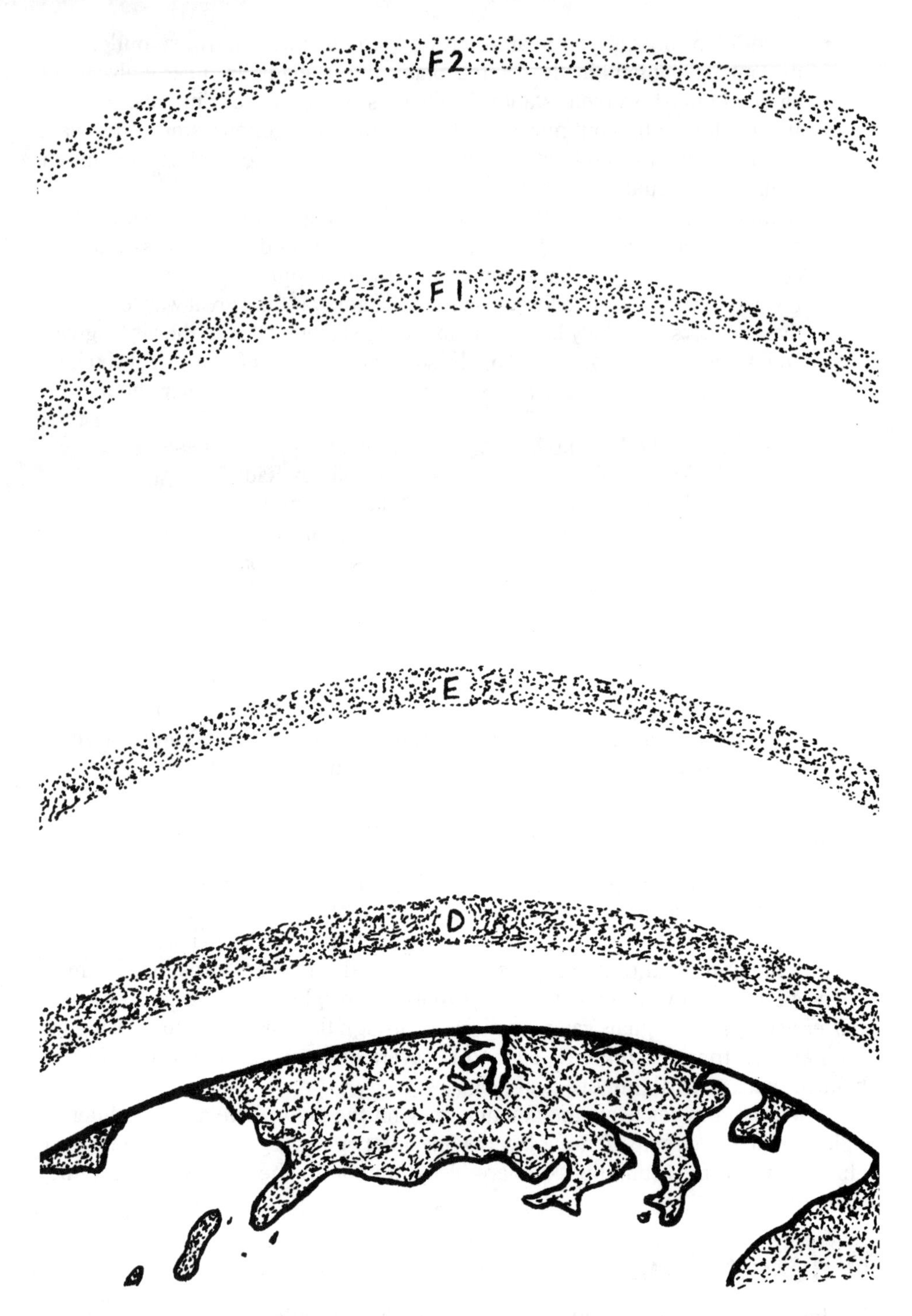

3-5 The earth is surrounded by several layers of electrically charged particles known as the *ionosphere*.

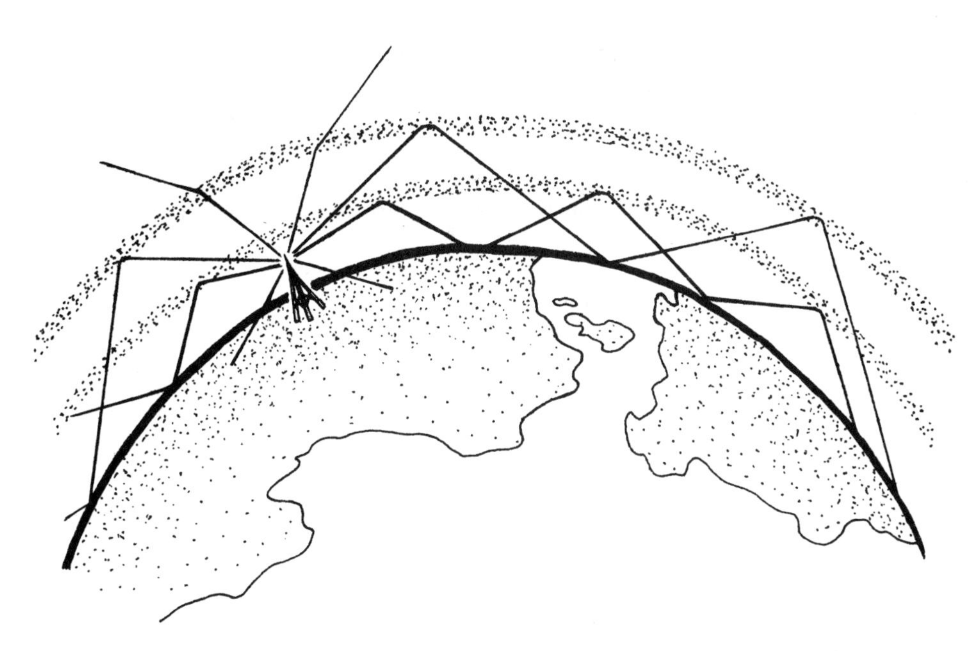

3-6 Each layer of the ionosphere has a different effect on radio skip conditions.

the bands are reserved for a variety of services, such as ship-to-shore communications, aviation radio, ham radio, military communications, and so on. The established broadcasting bands are:

3.200–3.400 MHz	Night band. Used mostly for domestic tropical shortwave services.
3.900–4.000 MHz	Night band. Domestic, mostly tropical shortwave stations. Shared with ham radio operators.
4.750–5.760 MHz	Night band. Domestic shortwave stations.
5.950–6.200 MHz	Night band. Domestic and international shortwave stations.
7.100–7.300 MHz	Night band. International shortwave, shared with ham radio operators.
9.500–9.900 MHz	All day, but more active at night. International shortwave.
11.650–12.050 MHz	All day. International shortwave.
13.600–13.800 MHz	All day. International shortwave.
15.100–15.600 MHz	All day. International shortwave.
17.550–17.900 MHz	All day, but more active in daylight hours. International shortwave.
21.450–21.850 MHz	Daytime. International shortwave.
25.670–26.100 MHz	Daytime. International shortwave.

The majority of shortwave stations operate within these designated frequency limits, which are sometimes shared with other services, such as ham radio and utilities. Because of overcrowded band conditions in some parts of the world, several international broadcast stations choose to operate slightly beyond the normal frequency range. Table 3-1 is a list of shortwave frequencies used for English (as well as other language) broadcasts.

Table 3-1. English language shortwave broadcast frequencies.

Station/country	Frequency
Albania	9.760, 11.825
RAE, Argentina	11.710
Australia	6.080, 7.140, 7.240, 9.580, 9.710, 11.880, 11.910, 11.930, 11.715, 11.745, 12.000, 13.605, 15.110, 15.160, 15.670
ORF/Austria	9.875, 13.730
BBC/England	5.965, 5.975, 6.175, 7.325, 9.515, 9.590, 9.740, 9.915, 11.750, 11.775, 12.095, 15.260
BRT/Belgium	9.925, 13.655, 13.710
R. Sofia/Bulgaria	9.595, 9.700, 11.680, 11.950
RCI/Canada	5.960, 9.535, 9.625, 9.755, 11.820, 11.845, 11.940, 13.720, 15.150, 17.820
R. Beijing, China	9.770, 11.715
R. Nacional, Colombia	11.822, 17.865
RFPI, Costa Rica	7.375, 15.030, 21.465
R. Habana/Cuba	11.760, 11.950, 13.700
Czechoslovakia	7.345, 9.540, 11.990
HCJB/Ecuador	9.745, 11.925, 15.115, 17.890, 21.455
R. Cairo, Egypt	9.475, 9.675, 17.595
Finland	9.560, 11.755, 15.400, 21.550
Deutsche Welle, Germany	6.045, 6.035, 6.055, 6.085, 6.145, 7.285, 9.515, 9.565, 9.610, 9.640, 9.770, 11.685, 11.945, 12.055
Ghana	6.130
Voice of Greece	7.430, 9.420, 11.645, 15.565, 15.650, 17.515
Guatemala	3.300
Mongolia	7.260, 13.780
R. Nederlands/Holland	6.020, 6.165, 11.835, 15.570
R. Budapest, Hungary	6.110, 9.835, 11.910
All India Radio	9.535, 9.910, 11.715, 11.745, 15.110
Iran	9.022, 9.765, 15.260
R. Baghdad, Iraq	11.830, 15.140, 15.455
KOL Israel	7.465, 9.435, 11.587, 11.605, 11.675, 15.640, 17.545, 17.590
RAI, Italy	9.575, 11.800
R. Japan	5.960, 11.815, 11.870, 15.430, 17.810, 17.835, 17.845
Kenya	4.935
Laos	7.112
Lithuania	9.870, 15.180, 17.065, 17.690
Luxembourg	15.350
Norway	11.925
Portugal	9.555, 9.600, 9.705, 11.840
R. Romania Int.	5.990, 6.155, 9.510, 9.570, 11.830, 11.940
R. Moscow/Russia	6.000, 6.045, 7.115, 7.150, 9.625, 9.715, 9.725, 11.762, 11.840, 12.050, 15.485, 17.700, 17.720, 17.890, 21.480, 21.685

Table 3-1. Continued.

Station/country	Frequency
TWR, Netherlands Ant.	9.535, 11.930
New Zealand	9.700, 15.120, 17.770
Nigeria	7.225
North Korea	6.576, 9.977, 11.335
Norway	9.645, 11.925
Poland	6.135, 7.270, 7.275, 9.525
R. Seoul, S. Korea	7.275, 9.640, 9.750, 11.805, 15.575
RSA/South Africa	11.900
Spanish National Radio	9.530
R. Sweden	9.695, 11.705, 17.870, 21.500
Swiss Radio Int.	6.135, 9.650, 9.885, 12.035, 17.730
Voice of Free China, Tiawan	5.950, 9.680, 9.765, 11.740, 11.860 15.345
Turkey	9.445
R. Kiev, Ukraine	7.400, 9.800, 15.180, 17.065, 17.690
R. Tashkent, Uzbekhistan	5.930, 5.995, 7.190, 7.265
Christian Science/USA	7.395, 9.455, 9.530, 9.850, 13.625, 13.760, 15.665, 17.555, 17.865, 21.670
KGEI/USA	15.280, 17.750
KLNS/USA	7.355
KTBN/USA	7.510, 15.590
KVOH/USA	17.775
WHRI/USA	7.315, 9.465, 9.495, 15.105, 21.840
WINB/USA	15.145
WMLK/USA	9.465
WRNO/USA	7.335, 15.420
WWCR/USA	7.435, 5.935, 12.160, 15.690, 17.535
WYFR/USA	5.950, 5.985, 7.355, 9.705, 11.830, 15.335, 17.760. 21.525. 21.615
Voice of America	5.995, 6.130, 7.405, 9.455, 9.775, 11.580, 15.120, 15.205, 17.735, 21.550
Vatican Radio	6.025, 6.095, 7.250, 7.305, 9.605
Venezuela	9.540
Yugoslavia	9.580, 17.740

International stations are always on the move

When an AM station is at 800 kHz (80 on the AM dial) in the morning, it will most certainly be there in the afternoon as well. You know exactly where to look when you want to hear the news, a ball game, a talk show, or some great music.

But shortwave stations are dealing with an entirely different situation. Instead of serving listeners living in one location—as AM and FM stations do—a shortwave station's audience is scattered all over the globe. And the "skip" conditions shortwave broadcasters rely on to transport their signal from one part of the world to another are always changing.

Everything from the time of day to the season of the year and the number of spots on the sun affects shortwave skip. The frequency that does a great job of delivering programs to a targeted area in the morning can do a very poor job that same evening.

To make the best use of the ionospheric skip conditions our atmosphere offers, the vast majority of international stations move (at prescheduled times) throughout the day from one frequency band to another. At first, this can make it challenging to keep up with your favorite programs. But overseas stations want you to hear them (that's why they're on the air!), and they do what they can to help. They print schedules of their frequencies and the times of day they use them. Broadcast schedules are sent out to listeners on request.

Most international broadcast stations have computerized mailing lists of listeners that want to receive schedules as soon as they come out. Ask to have your name added, and you'll always be up-to-date on when and where to tune in. Monthly magazines such as *Monitoring Times* and *Popular Communications*, radio clubs, and annual books (*Passport to World Band Radio* and *The World Radio-TV Handbook*) can help you keep track of broadcast times and frequencies. They are excellent sources of information for both new and experienced listeners.

Relay stations

A number of international broadcasters use relay stations to improve signal strength and quality in remote parts of the world. The first step in using a relay station is to send the signal up to a satellite that's in orbit 22,300 miles above us. The satellite beams it back down to earth. The dish antenna at the relay station picks it up, amplifies it, and rebroadcasts the program on the shortwave frequency band (Fig. 3-7).

Islands in the Caribbean and Canadian locations are favorite relay sites for delivering programs to North America. Many relay stations are built to receive and rebroadcast only one station's programs. But relay services are also provided by international shortwave broadcasters, sometimes in exchange for airtime on the other station's transmitters.

Here are a few examples of how relay stations are used to better reach listeners in North, South, and Central America:

- Radio Nederlands relays via The Netherlands Antilles (Caribbean).
- The British Broadcasting Corporation and Deutsche Welle (Germany) relay via Antigua (Caribbean).
- The British Broadcasting Corporation, Deutsche Welle, Radio Japan, Radio Austria International, Radio Beijing, and Radio Korea all relay via the Sackville facility in Canada.

Figure 3-8 is the Voice of America's relay station in Rhodes, Greece.

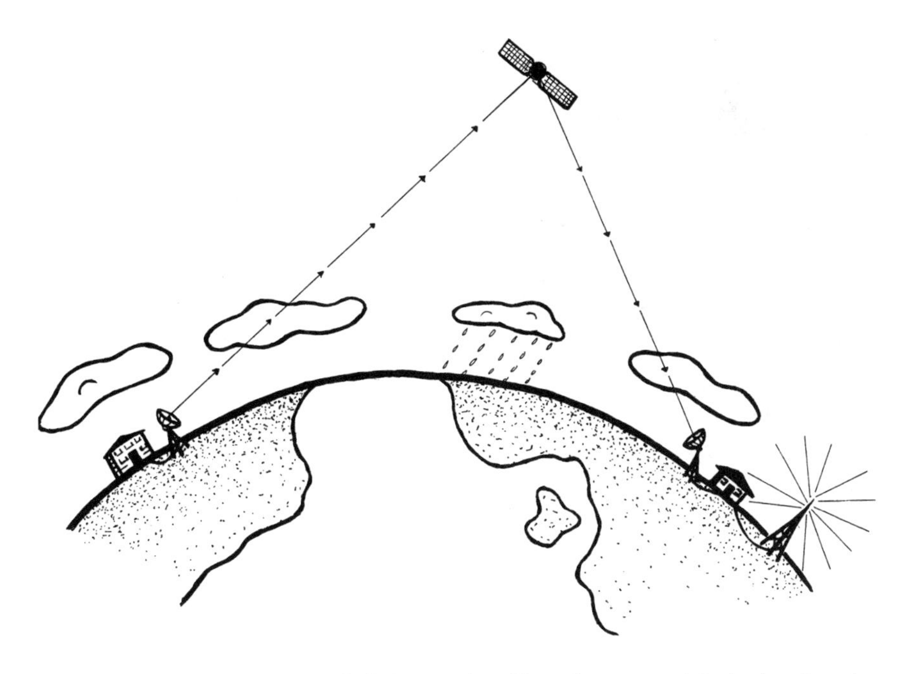

3-7 Relay stations help international broadcasters get their signals out.

Interval signals help you get on frequency

Just before they start their regularly scheduled broadcast, most international shortwave broadcast stations transmit an interval signal. Interval signals are made up of tones or musical notes, repeated over and over again. In addition to sign-on interval signals, some stations play interval signals for a few minutes between various language services. It helps listeners to locate the station so they won't miss out on the first few minutes of the program while they're trying to get on the right frequency. Of course, it's easy to tell which frequency you're on when the radio you own has digital readout. But interval signals are extremely helpful if you're using a receiver with an analog dial.

Domestic service transmissions

Since their broadcasts are intended for a local or regional audience, domestic shortwave stations are usually low-powered and are almost always in the language of the local people. Some are government-owned and some are privately owned. Often, they operate in a fashion similar to our own AM/FM stations—broadcasting programs of local news, weather, commercials, announcements, and of course, the popular music of their part of the world.

In a number of third-world countries, domestic shortwave stations are used to educate a widely scattered and often illiterate population in skills that can help improve their daily lives. Unless you live in their intended reception area, domestic ser-

3-8 The VOA relay station in Rhodes, Greece. VOA photo.

vice broadcasts can be very difficult to hear. But under the right skip conditions, you can have surprisingly good reception of even these low-power outlets.

Domestic programming on high-power transmitters

A few international shortwave stations are making it easier to keep up-to-date on the local angle of what's going on in their country, and save on expenses at the same time.

Instead of producing separate programs for overseas listeners, they fill at least part of their daily schedule with high-powered relays of domestic service transmissions.

Radio New Zealand International, for one, serves as a relay for their country's domestic "National Radio" transmissions. You'll get to hear local news, weather, sports, music (both popular and classical), talk shows, announcements of community events. You'll even know when there's a major water pipe break or traffic tie-up in the city of Wellington.

Pirate and clandestine stations

If you know where to tune on your shortwave radio, you're likely to run onto some stations that, for one reason or another, have decided to broadcast without a license.

Pirates of the airwaves

The most popular category of unlicensed station on the shortwave bands are pirate stations, also known as *free radio* stations and hobby broadcasters. Pirate station operators risk government fines and confiscation of equipment in order to gain access to the public airwaves.

Since pirates want to avoid being tracked down by the government, their broadcast schedule is very erratic. Some pirate radio operators broadcast only once or twice a year (usually on a holiday), while other pirates switch on their transmitters every week or two. But as a general rule, the best time to try for a pirate is on weekends between sundown and midnight. Figure 3-9 is a card from holiday (Christmas/New Year's Eve) AM pirate WJDI.

Although they can show up anywhere, pirates usually appear on or around 6.2–6.3 MHz (evenings), 7.415–7.420 MHz (evenings) and 15.045 MHZ (late afternoons). From time to time, they can also be heard just above the AM band–1620 to 1700 kHz.

"Free radio" shows are always innovative, exciting, and entertaining. Unlike the predictable format many stations follow, pirates specialize in the unexpected. Their shows are whatever the station operator decides they should be that night. They often consist of spirited conversation between announcers, rock music, and hilarious comedy skits, many of which are written and performed by station operators.

Writing to pirates

Pirate radio operators always enjoy getting fan mail from their listeners. They want to know who you are, how you liked their program, and how well their signal came into your area. But no smart operator would want to reveal his true location to any government agent that might be listening. So instead of using their home address, pirates give the address of a mail drop—an individual who receives their mail and forwards it on to them.

To cover the extra postage expenses involved in all this remailing (most pirates and mail drop operators aren't rich!), listeners are asked to enclose three first-class stamps when they write to a pirate station. The first stamp gets it from the mail drop to the pirate, the second gets the reply from the pirate back to the mail drop, and the third stamp takes the pirate's letter from the mail drop to the listener.

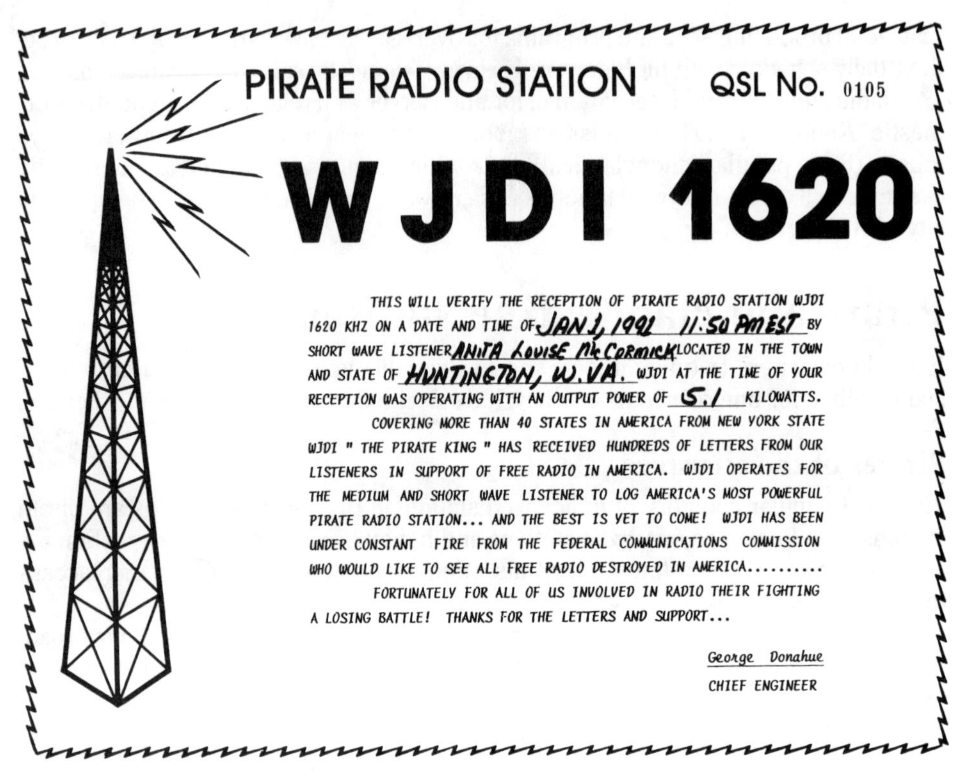

PIRATE RADIO STATION QSL No. 0105

WJDI 1620

THIS WILL VERIFY THE RECEPTION OF PIRATE RADIO STATION WJDI 1620 KHZ ON A DATE AND TIME OF JAN 1, 1991 11:50 PM EST BY SHORT WAVE LISTENER ANITA LOUISE McCORMICK LOCATED IN THE TOWN AND STATE OF HUNTINGTON, W.VA. WJDI AT THE TIME OF YOUR RECEPTION WAS OPERATING WITH AN OUTPUT POWER OF 5.1 KILOWATTS.

COVERING MORE THAN 40 STATES IN AMERICA FROM NEW YORK STATE WJDI " THE PIRATE KING " HAS RECEIVED HUNDREDS OF LETTERS FROM OUR LISTENERS IN SUPPORT OF FREE RADIO IN AMERICA. WJDI OPERATES FOR THE MEDIUM AND SHORT WAVE LISTENER TO LOG AMERICA'S MOST POWERFUL PIRATE RADIO STATION... AND THE BEST IS YET TO COME! WJDI HAS BEEN UNDER CONSTANT FIRE FROM THE FEDERAL COMMUNICATIONS COMMISSION WHO WOULD LIKE TO SEE ALL FREE RADIO DESTROYED IN AMERICA..........

FORTUNATELY FOR ALL OF US INVOLVED IN RADIO THEIR FIGHTING A LOSING BATTLE! THANKS FOR THE LETTERS AND SUPPORT...

George Donahue
CHIEF ENGINEER

3-9 A card from AM pirate station WJDI.

Figure 3-10 maps out how the system works. First, your letter goes to the mail drop operator's address. He takes it home, repackages it, and forwards it to the pirate. When the pirate receives your letter, he fills out one of his station cards, puts it in an envelope, and sends it back to the drop—who, in turn, forwards it to you. Currently, the most active mail drops are:

- PO Box 109
  Blue Ridge Summit, PA 17214
- PO Box 452
  Wellsville, NY 14895

While it is against the law to operate a pirate station, there is certainly no regulation against writing to one. In fact, tuning in pirate stations, corresponding with them, and collecting the souvenir cards they send can be among the most exciting aspects of your shortwave listening hobby. Figure 3-11 shows two hand-drawn cards from North American pirate stations.

Why do they do it?

With current regulations, the costs involved in owning and operating even a small radio station on the AM, FM, or shortwave bands far exceeds what most people can ever hope to afford. Pirates believe that the average person deserves access to the

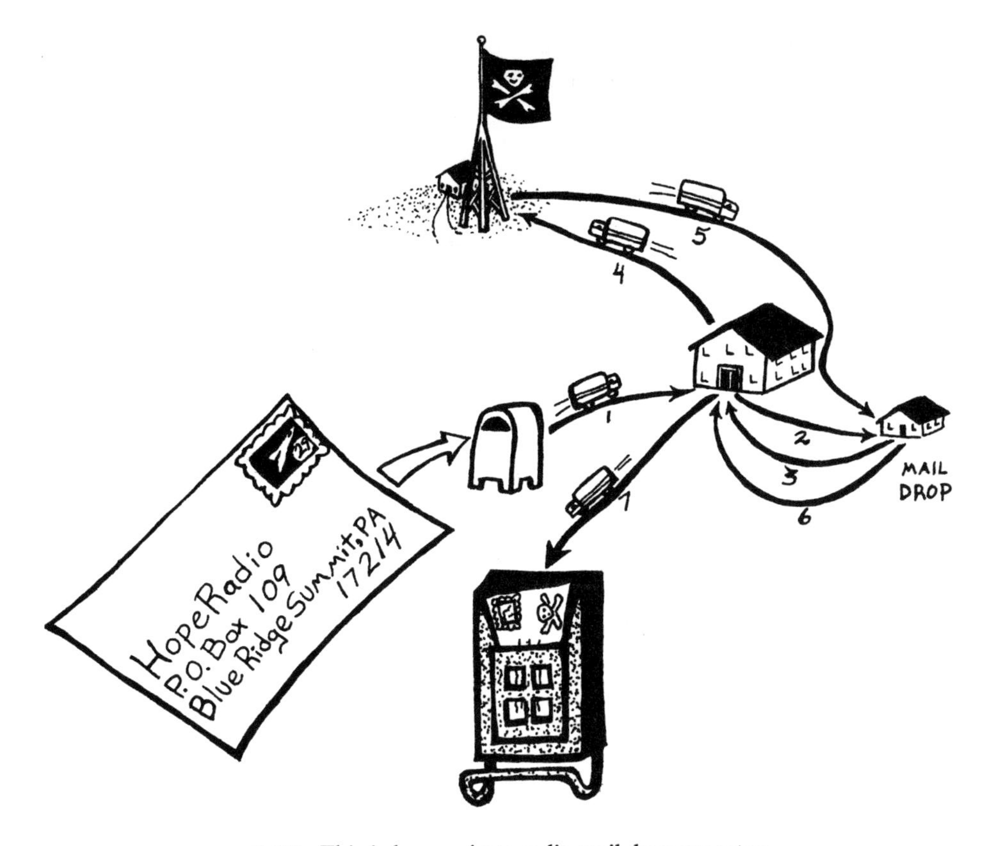

3-10 This is how a pirate radio mail drop operates.

airwaves. Pirates go on the air without a license to show the public, as well as the government, that small, low-cost, low-power stations do serve a purpose that isn't being fulfilled by existing commercial broadcast stations. In fact, many pirates say that they would be happy to "go legal" if the Federal Communications Commission were to issue a hobby broadcast license with reasonable cost and equipment requirements.

New legal ways for pirates to broadcast

In the meantime, several groups of free radio operators have found new and innovative ways to put their shows on the air without having to worry about that dreaded knock on the door when the FCC finally catches up with them. WWCR, a 100,000-watt shortwave station in Nashville, Tennessee, has been selling time to Radio Newyork International since the fall of 1990. Al Weiner, Pirate Joe, Johnny Lightning, Big Steve Cole, and their friends have a large and faithful listening audience. These former pirates, who had attempted to operate an offshore station in international waters a few years earlier, play music, take calls from listeners, read letters from free radio enthusiasts around the world, and discuss important issues of the day (Fig. 3-12). They can be heard from 10 P.M. to 1 A.M. Eastern Time (7 to 10 P.M. Pacific) Sunday nights on 7.435 MHz.

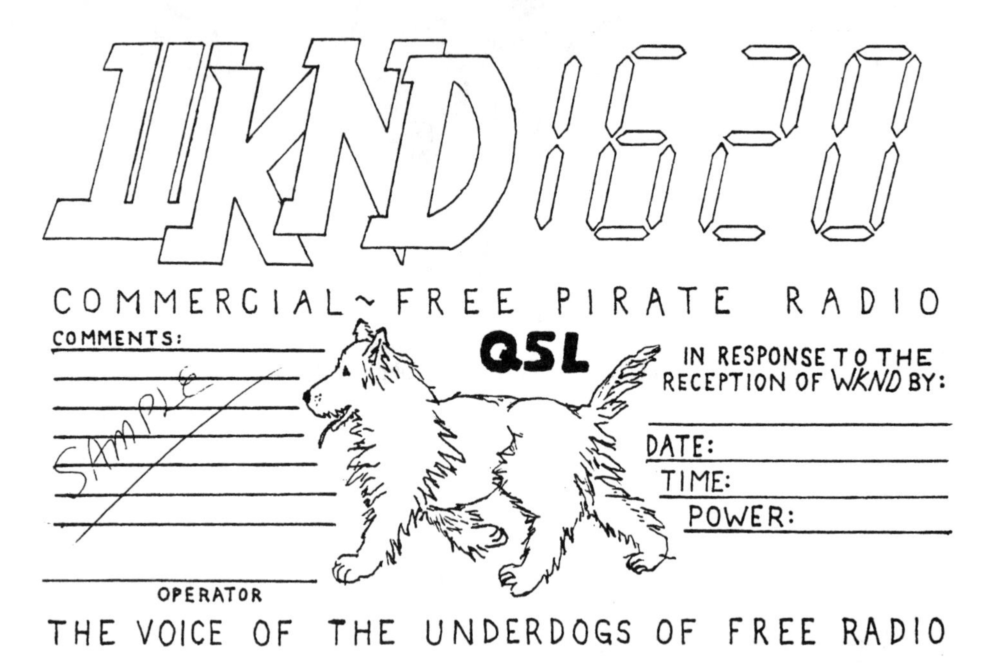

3-11 Here are cards from pirate station Hope Radio International and WKND, "The Voice of the Underdogs of Free Radio."

The Radio Newyork International crew pays for their Sunday nighttime slot by selling RNI T-shirts and videos, airing commercials (mostly for radio-related products), and asking for donations from their listeners.

Within a few months of RNI's appearance on WWCR, another group of former pirates joined the Nashville station's lineup as Radio Free New York. Their listeners could write in and receive not only their current QSL card (Fig. 3-13), but cards from their past stations (Fig. 3-14) as well!

RADIO NEW YORK INTERNATIONAL

CERTIFICATE OF APPRECIATION

Presented To

Anita Louise McCormick

for your support of free radio and

RADIO NEW YORK INTERNATIONAL

Thanks for Being such a good listener!!

Feb. 7th, 1991

Date

Allan H. Weiner, President

RADIO NEW YORK INTERNATIONAL

3-12 A Radio Newyork International OSL—and a RNI Certificate of Appreciation, which goes out to listeners that buy RNI merchandise or contribute to the station.

Dear RFNY Listener:

Radio Free New York operates every Saturday via WWCR, Nashville on a frequency of 7520 kilocycles from 0400 to 0700 UTC.

This confirms your reception report of RFNY from ______________ at ______________.

Thanks again, and feel free to stay in touch with your questions and comments about RFNY! Keep listening to the ONLY voice from the HEART of New York City -- Radio Free New York!!!

3-13 This great card came from Radio Free New York. John Calabro.

3-14 Pirate station WHOT, AM and FM. John Calabro.

A few existing pirate stations and would-be pirates have recently signed up to buy airtime on the Becker Satellite Network. Anyone that has a backyard satellite dish can pick up their signal. With the FCC pushing for higher fines for unlicensed broadcasting, and prices falling on satellite dish equipment, this could well be the trend of free radio broadcasting in the future.

Want to know more?

You can learn more about these fascinating stations by reading:

Pirate Radio Stations—Tuning in Underground Radio Broadcasts, by Andrew Yoder (TAB Books).
The Pirate Radio Directory, by George Zeller (Tiare Publications).
The Complete Manual of Pirate Radio, by Zeke Teflox (Acme Enterprises).
Offshore Echoes (catalog/magazine—covers offshore and European pirate activity), PO Box 1514, London W72LL England.

Clandestine stations—revolution by radio

Clandestine stations, by contrast, use their airtime almost exclusively for getting across their political positions. These are very common in areas where government upheaval is taking place. If a revolution is in the making, a clandestine station is often there to urge it on. They can show up on almost any frequency and any part of the band.

Some clandestine stations operate in a "safe" territory and are able to keep a regular schedule. Others, who operate in unfriendly territory, try to avoid government detection by making short, irregular transmissions. Nearly all clandestine stations broadcast only in the language of the intended audience.

Keeping up with the latest developments

One of the best sources of current information on pirate and clandestine radio activity is the Association of Clandestine Enthusiasts, known as the A*C*E*, PO Box 11201, Shawnee Mission, KS 66270-0201. They publish a club bulletin (Fig. 3-15) of monthly columns, members' station loggings, and general information on the underground radio scene.

Another good source of information is the biweekly publication, *The Pirate Pages,* PO Box 109, Blue Ridge Summit, PA 17214.

Monitoring Times and *Popular Communications* also feature monthly columns of pirate and clandestine information.

Hearing ham radio operators

Amateur radio operators from around the world communicate on the shortwave bands. While shortwave broadcast stations are required by law to stay on preapproved frequencies, ham radio operators are free to move around and use any frequency they wish, as long as they stay within their designated frequency bands.

Most of the hams you hear on shortwave communicate with either International Morse Code or single sideband—a method of transmitting voice that doesn't take up as much band space as the signals international shortwave broadcast stations put out. (You'll read more about this in later chapters.)

If you want to listen to single sideband transmissions, you'll need to buy a shortwave radio with a BFO or SSB control. Otherwise, SSB transmissions come out of your speaker as unintelligible garble. Even if your radio doesn't receive SSB, you can still hear a fair amount of ham activity.

The following frequencies are often used by hams that like to operate older tube-type radios and transmit in good old-fashioned AM:

Night MHz	Day MHz
3.870	14.285
3.880	14.295
3.890	21.385
	29.00–29.20
	28.305–28.325

Many of today's hams use computers to communicate. With a modemlike device connected between your computer and your ham radio gear, you can send and exchange information with any other ham radio operator in the world that owns a similar setup.

3-15 A*C*E* keeps members up-to-date on pirate and clandestine activity. Courtesy of the A*C*E*.

Chapter 7 tells you how to go about getting your license and putting a ham radio station of your own on the air.

Utility stations

In the world of radio, the term "utility station" covers a wide variety of transmissions. Any station that doesn't broadcast directly to the public and isn't part of the ham radio service is classified as a utility. The category includes ship-to-shore transmissions, overseas phone calls, weather forecasts, aviation communications (mostly overseas flights), barges on the river, military communications, and so on.

Utility stations use voice, Morse code, radioteletype, and computer code (ASCII) to get their messages across. Nearly all utility voice transmissions are in single sideband; so if you plan to hear them, make sure that the shortwave radio you buy is equipped with a BFO/SSB control.

Figure 3-16 is a listing of popular frequencies from *Monitoring Times*'s monthly Utility World column. When you check these frequencies, remember that utility stations operate for only short periods—when they have messages to transmit—and are silent most of the time. If you don't hear something on a favorite frequency, try again later. Or better yet, explore some new frequencies. You never know what you might hear! Table 3-2 is a list of abbreviations you're likely to encounter in utility listening.

Decoding digital transmissions

More and more utility stations are going high-tech. They communicate in RTTY (radio teletype), ASCII computer language, Morse code, and other digital modes that the average shortwave receiver is unable to translate. If you want to know what all those dits, dahs, chirps, and hums between shortwave stations are really trying to say, multimode decoders can be an exciting addition to your listening hobby.

The MicroDec, manufactured by Somerset Electronics, Inc., is one of the most advanced, yet easy-to-use decoders on the market. It brings in news services from around the world, government message traffic, maritime, military, and amateur radio digital communications (Fig. 3-17).

Once digital signals are decoded, they can be read directly on the unit's eight-character LED dot matrix data display. With the VIP50 Video Printer Interface, you can display the decoded information on your TV screen, or if you own a computer, you can display the information on your monitor and interface to a parallel dot matrix printer. The MicroDec Series has a standard serial data output that interfaces with virtually any personal computer terminal. But you really don't need a computer for storage and retrieval. Simply use your VCR with the VIP50 and videotape the decoded information.

Tuners are available from Somerset Electronics that permit you to vary the bandwidth to better receive Morse code, teletype, ASCII, and other digital transmissions. Tuners also enable you to use a MicroDec to decode transmissions from your scanner.

The MicroDec is lightweight, rugged, and battery-powered. With the recharge-

MONITORING TIMES UTILITY WORLD HOT 160 FREQUENCIES

2182 Intl Emergency
2638 USCG Broadcast
2670 USCG Broadcast
2716 USN Harbor Common
3023 USN Tactical
3067 USAF A/G channel
3081 USAF A/G channel
3123 USCG/USN aircraft
3130 USN tactical
3144 USAF A/G channel
4426 USCG weather
4466 Civil Air Patrol
4469 Civil Air Patrol
4506 Civil Air Patrol
4509 Civil Air Patrol
4517 USAF MARS
4582 CAP Emergency
4593.5 USAF MARS
4637.5 Offshore Petro
4670 Spy numbers
4675 International ATC
4722 RAF Flight weather
4727 SAC Victor/S-390
4742 Spy numbers
4746 USAF A/G channel
4747 USAF A/g channel
5000 World Time/Freq
5015 Corps of Engineers
5302 USAF A/G channel
5320 US Coast Guard
5520 International ATC
5550 International ATC
5598 International ATC
5616 International ATC
5649 International ATC
5680 Search & Rescue
5688 USAF A/G channel
5692 US Coast Guard A/G
5696 US Coast Guard A/G
5703 USAF tactical A/G
5710 USAF A/G channel
5812 Spy Numbers
6501 USCG weather
6577 International ATC
6586 International ATC
6604 VOLMET weather
6673 Hurricane Hunters
6683 Mystic Star
6697 USN Hicom channel
6705 USAF A/G channel
6723 USN Raspberry
6738 USAF A/G channel
6750 USAF A/G channel
6753 Canforce channel
6756 Mystic Star
6757 USAF A/G channel
6761 SAC A/G Quebec/S391
6802 Spy numbers
6840 Spy numbers
6927 Mystic Star
7335 Canadian Time Stn
7527 US Customs Service
7635 CAP Primary
7917 Civil Air Patrol
8764 USCG weather
8825 International ATC
8846 International ATC
8864 International ATC
8879 International ATC
8891 International ATC
8918 International ATC
8964 USAF A/G channel
8967 USAF A/G channel
8972 USN Safe for Flight
8984 US Coast Guard A/G
8989 USAF A/G channel
8993 USAF A/G channel
9011 USAF A/G channel
9014 USAF A/G channel
9018 Mystic Star
9027 SAC Romeo/S-392
10000 World Time/Freq
10051 VOLMET weather
10493 FEMA channel
10780 NASA A/G channel
11176 USAF A/G channel
11179 USAF A/G channel
11200 RAF Flight weather
11207 USAF A/G channel
11214 USAF A/G channel
11226 USAF A/G channel
11228 USAF A/G channel
11233 Canforce channel
11234 RAF A/G channel
11236 USAF A/G channel
11239 USAF A/G channel
11243 SAC Alpha/S-393
11246 USAF A/G channel
11267 USN Hicom channel
11279 International ATC
11282 International ATC
11300 International ATC
11309 International ATC
11387 International ATC
11396 International ATC
11398 Hurricane Hunters
11494 SAC A/G Lima/S-311
12047.5 USN MARS Afloat
13089 USCG weather
13181 USN Hicom channel
13201 USAF A/G channel
13214 USAF A/G channel
13215 USAF A/G channel
13241 SAC A/G S-394
13244 USAF A/G channel
13247 SAC-TAC W-109
13270 VOLMET weather
13282 VOLMET weather
13291 International ATC
13306 International ATC
13354 Hurricane Hunters
13826 USN MARS channel
13950 USMAG-Latin America
13974 USN MARS channel
14313 Ham Maritime Net
14325 Ham Hurricane Net
14384.5 CFARS (Canada)
14441.5 USN MARS calling
14470.0 USN MARS channel
14606 USAF MARS channel
14686 DEA channel
14902 USAF MARS/FEMA
15000 World Time/Freq
15014 USAF A/G channel
15015 USAF A/G channel
15031 USAF A/G channel
15025 Canforce channel
15036 USAF A/G channel
15041 SAC A/G channel Mike
15522 USN Hicom channel
16454 USAF A/G channel
17904 International ATC
17907 International ATC
17946 International ATC
17975 SAC A/G Tango/S-395
18002 USAF A/G channel
18009 USN Hicom channel
18019 USAF A/G channel
18027 Canforce channel
18666 DEA channel
20000 World Time/Freq
20192 Space shuttle audio
20631 SAC A/G Whiskey
20885 USMAG-Latin America
21937 Hurricane Hunters
23227 USAF A/G channel
23287 USN Hicom channel
23315 USN Hicom channel
23337 SAC A/G Uniform
23403 DEA channel

3-16 Utility World Top 160 Frequency List. Monitoring Times.

Table 3-2. Common utility abbreviations.

AFB	Air Force Base
ASCII	American Standard Code for Information Interchange (most commonly used code for exchanging computer data)
ATC	Air Traffic Control
CAP	Civil Air Patrol
CG	Coast Guard
CQ	General call any station may answer
CW	Continuous wave—Morse code
DE	French word "from," used in CW and other digital communications
DOT	Department of Transportation
EAM	Emergency Action Message
FAA	Federal Aviation Administration
FEMA	Federal Emergency Management Agency
ID	Identification
LSB	Lower Side Band
MARS	Military Affiliate Radio System
M/V	Motor Vessel
Net	Network
RAF	Royal Air Force
RTTY	Radio Teletype
SAC	Strategic Air Command
SAR	Search and Rescue
TAC	Tactical
UNID	Unidentified
USAF	U.S. Air Force
USB	Upper Side Band
USCG	U.S. Coast Guard
USN	U.S. Navy

3-17 MicroDec Multi-Mode Decoder and a VIP50 Video-Printer Interface. Somerset Electronics, Inc.

able battery option, it is the only self-powered, portable multimode decoder available on the market. If you have a portable shortwave receiver and would like to explore the world of digital transmissions, the MicroDec would make an excellent accessory. MicroDec products are available from:

Somerset Electronics, Inc.
1290 Hwy A1A
Satellite Beach, FL 32937
(800) 678-7388

Shortwave radio on cable TV

C-Span is famous for bringing congressional and senate hearings into your livingroom. And now it is introducing millions of people to the exciting world of shortwave radio.

If your cable TV company carries C-Span, find out if it also carries the C-Span Audio Networks. Marketed as "A Composition of Global Voices in a Language You Can Understand," these networks truly live up to their billing. You can hear news, comment, and music from shortwave stations in every part of our planet—England, Germany, China, Japan, Austria, Canada, Korea, and Cuba. Even the United States' own Voice of America has a time slot.

If you haven't purchased a shortwave radio yet, tuning in the C-Span Audio Networks is a great way to experience international listening. C-Span 1 rebroadcasts English language programming from various shortwave stations around the world. C-Span 2 carries live retransmissions of the British Broadcasting Corporation's World Service programming 24 hours a day. If you have a satellite dish, you can pick up C-Span Audio Networks directly from Galaxy III—C-Span Transponder 24. C-Span Audio 1 is on the 5.22 audio subcarrier; C-Span Audio 2 is on the 5.40 audio subcarrier. For an up-to-date program schedule, write to:

C-Span Audio Networks
400 North Capitol Street, N.W.
Suite 650
Washington, DC 20001

Communicating with shortwave stations

Once you've had a chance to tune around on the shortwave bands and see what international radio is all about, you'll probably want to let your favorite stations know that you're out there listening. Shortwave stations love to hear from their listeners. They want to know who you are, how you liked their shows, and how well their signal made the trip into your area. And to encourage you to write in with your comments, they have some great promotional and souvenir items that are yours for the asking.

THANK YOU FOR YOUR RECEPTION REPORT
NOUS VOUS REMERCIONS POUR VOTRE RAPPORT D'ÉCOUTE
AGRADECEMOS SU INFORME DE RECEPCIÓN

11588 kHz Date/día 7.12.91 / 0500 GMT

YOUR REPORT HAS BEEN CHECKED AND AGREES WITH OUR LOG
VOTRE RAPPORT A ETE VERIFIE ET TROUVE CONFORME A NOS SPECIFICATIONS
SU INFORME HA SIDO VERIFICADO Y CONCUERDA CON NUESTROS DETALLES

3-18 A QSL card from KOL Israel.

QSL cards—what they are and how to get them

If you've been listening to shortwave radio for any length of time, you've probably heard of QSL cards and are wondering what they are, why people want them so bad, and what you have to do to receive one. To start with, QSL is a ham radio term for "I verify reception of your station." And that's exactly what a QSL card does! Radio sta-

3-19 A QSL card from The British Broadcasting Corporation.

tions mail them out to listeners that write in and let them know how well their signal is doing in their part of the world.

The front of a QSL card usually features a logo, design, or photo that has something to do with the station or country. Figure 3-18 is a QSL from KOL Israel. A QSL can show a photo of the station building, antenna array, or a picture of a popular announcer. It can also be a postcard scene of one of the area's popular tourist attractions, like the QSL card shown in Fig. 3-19 from the BBC, otherwise known as The British Broadcasting Corporation.

To receive a QSL card, you'll have to make out what is known in the world of radio as a reception report. In the report, you'll note the day and time you heard their program, the frequency you were tuned to (shortwave stations often broadcast on two or more frequencies at the same time), and a few details of what you heard on the program to prove that you were indeed listening to the station that you claim. And, of course, every station you report to will want to know just how well their broadcast can be heard in your area.

When you're ready to start writing in for QSL cards, chapter 5 tells you everything you'll need to know to make out the kind of accurate and detailed reception reports station engineers can use to aim a better signal into your part of the world. And not only that—you'll find out how you can get your letters read to a worldwide audience on international "Mailbag" programs! Almost every shortwave station has them—and nothing delights a mailbag host more than hearing from new listeners.

4
Station profiles

IN THIS CHAPTER, WE WILL TAKE A CLOSER LOOK AT SEVERAL OF THE STATIONS you're likely to hear on the shortwave bands.

The British Broadcasting Corporation

The British Broadcasting Corporation, commonly known as the BBC, has always been an important member of the shortwave broadcasting community. They have an excellent selection of news, entertainment, and feature programs, and a worldwide network of relay stations that make it easy to hear the BBC wherever you live.

For decades now, the BBC has been famous for their superb news coverage, not only of events in England, but of happenings around the world (Fig. 4-1). The BBC is known not only for news, but for its quality programming in all fields. If you're interested in literature, tune in the BBC's book review or short story programs. And you'll also enjoy its radio plays.

If music is your thing, you're sure to find some programs you like on the BBC. Rock fans will enjoy "Multi-track," "A Jolly Good Show," and others. Country fans will like "Country Style." Folk music fans won't want to miss "Folk in Britain." And even lovers of classical music and jazz have programs of their own in the BBC's lineup.

Listeners whose interests lie more on the spiritual side of life will want to hear "Words of Faith" and the various religious services and discussions aired on the BBC. Environmentalists won't want to miss "Global Concerns." And everyone that's interested in science will tune in "Discovery" on a regular basis. And even that is only part of the BBC's weekly lineup. There are farm programs, game shows, international sports, and the BBC communications program, "Waveguide." The BBC is easy to find. If you tune around a little, you should have no trouble finding the station broadcasting in English on nearly every international shortwave band.

4-1 BBC's John Eidinow, standing in front of Big Ben. BBC photo.

Radio Canada International

Radio Canada International has been a dependable friend to shortwave listeners around the world for many years. Their informal, yet professional style and their excellent variety of Canadian and world news, features, and music programs make them stand out from the crowd on the shortwave frequency bands.

Because of budget cutbacks, some of the programs currently heard on RCI are a relay of domestic programming. Other shows, by contrast, are produced especially for foreign listeners.

In addition to airing its own programming, Radio Canada International serves as a relay station for foreign stations that would otherwise have difficulty getting a decent signal into North America. Stations relayed from Canada include the BBC, Deutsche Welle, Radio Japan, and Radio Austria International.

Figure 4-2 shows some of the attractive QSL cards issued by Radio Canada International over the years.

4-2 An attractive display of QSL cards from Radio Canada International. RCI photo.

You can find RCI broadcasting in English on a number of shortwave frequencies: 5.960, 9.535, 9.625, 9.755, 11.845, 15.150, and 11.820 MHz.

Radio for Peace International, Costa Rica

Radio for Peace International is the only station of its type on the planet. Their focus is exclusively on peace, ecology, and social justice issues. Their studios and broadcasting facilities are located on United Nations land at the University for Peace in Costa Rica.

RFPI has been transmitting their message of peace and understanding to the world (with occasional disruptions due to earthquakes and tropical storms) since 1987. Frequencies include 7.375, 13.630, 15.030, and 21.565 MHz. Since it is nonprofit and noncommercial, Radio for Peace International depends on grants from supporting organizations and the generosity of their listeners for operating revenue. The station encourages people to make donations by offering souvenir items, such as RFPI T-shirts and a quarterly newsletter, to contributors. Figure 4-3 is an RFPI bumper sticker.

RADIO FOR PEACE INTERNATIONAL

Ciudad Colon, Costa Rica

7.375 13.630 15.030 21.465

Global community radio dedicated to peace.

4-3 RFPI's bumper sticker.

Radio for Peace International's programming is as diverse and intriguing as the world we live in. Depending on when you tune in, you might hear the rich and exotic music of a little-known tribal society on "Sound Currents of the Earth," a report on the treatment of political prisoners on "Amnesty International Reports," a guest speaker from almost anywhere on "World Citizen's Hour," commentary and music from Al Weiner, Pirate Joe, and their friends on "Radio Newyork International," or the latest news about international communications and broadcasting on Glenn Hauser's "World of Radio."

RFPI's signal currently reaches an audience of more than 35,000 people in over 50 countries. And with a new, more powerful transmitter under construction, its audience is sure to grow even more in the future.

Deutsche Welle—The Voice of Germany

When the Berlin Wall fell, shortwave listeners around the world tuned to Deutsche Welle to keep up with the moment-by-moment developments. East and West Germany, divided by the wall for decades, had started the process of becoming one nation again.

Communist regimes no longer control Eastern Europe. Still, Germany's problems are far from over. Taking down a wall is one thing. Merging two very different economic systems is something else. But no matter what happens, you can depend on Deutsche Welle, The Voice of Germany, to keep shortwave listeners everywhere informed.

When people tune into Deutsche Welle, they can depend on a lively blend of news, editorials, music, and features. Their lineup includes "European Journal," "Living In Germany," "Through German Eyes," "Science and Technology," "Religion and Society," "Man and the Environment," and a mailbag program where questions and comments from listeners are read. Figure 4-4 is a photo of the Deutsche Welle North American Service Staff.

4-4 Deutsche Welle's North American staff.

If you write in and request it, you can be put on the mailing list to receive Deutsche Welle's free newsletter, *Tune In*. And while you're at it, why not request a QSL card (Fig. 4-5)? Deutsche Welle's frequencies for English transmissions include: 6.045, 6.035, 6.055, 6.085, 6.145, 7.285, 9.515, 9.565, 9.610, 11.685, and 11.945.

HCJB—Quito, Ecuador

HCJB, "The Voice of the Andes," has been a mainstay of shortwave listening for many years now. In fact, they recently celebrated their 60th anniversary. While HCJB is primarily an evangelical Christian station, you can also find a number of other interesting programs in their schedule.

To start with, they play beautiful Ecuadorian music you're not likely to hear anywhere else. They broadcast historical programs, shows about Ecuadorian nature and culture, and news about Ecuador, South, and Central America. HCJB produces "DX Partyline," with shortwave listening tips from around the world, and presents a program about ham radio for the amateurs in their audience.

Listeners are encouraged to write to HCJB and request one of their colorful QSL

4-5 A Deutsche Welle QSL card.

cards. A new design is printed six times a year. You can tune them in on: 9.745, 11.925, 15.155, 17.890, and 21.455 MHz.

Radio Moscow, Russia

For decades, Radio Moscow was the voice of worldwide communist propaganda. Now those same transmitters and studios are being put to a very different use. Even a few years ago, who would have believed that Radio Moscow would air a program such as "New Market," promoting foreign investment in Russian products and services? Or run commercials for privately owned business and industry? In Russia, as in all countries of the former Soviet Union, radio is changing.

At the time of this writing, Radio Moscow airs national and world news, business reports, music (both pop and classical), readings from Russian books, mailbag programs, and a number of other features such as special daily reports on Asia and the Pacific.

While Radio Moscow is still easy to hear in North America, it is no longer the radio superpower it once was before the breakup of the Union of Soviet Socialist Republics. Some of Radio Moscow's transmitters are now being leased by foreign stations—such as the BBC—to help improve their coverage of Asia. Other transmitters were handed over to former Soviet countries after they gained their independence.

Shortwave listening is much more common in Russia than it is in North America. Almost every household has at least one shortwave receiver. Even Gorbachev listened to foreign shortwave stations to keep up with developments when he was being held in captivity during the 1991 communist hard-liner uprising. Now that Russia is going into free-market capitalism, new, independently owned stations on the AM, FM, and shortwave bands are starting up all over the country.

Radio Nederlands

For decades, Radio Nederlands, the Dutch voice to the world, has been one of the most popular stations on the shortwave bands. Their "Newsline" program, heard at the beginning of each broadcast, keeps listeners up-to-date on events happening in Holland, Europe, and around the world.

"Media Network," hosted by Jonathan Marks, keeps you informed about events in all fields of communication, especially changes in programming, frequencies, and so on, in the shortwave frequency spectrum. Radio enthusiasts from all over the world write, FAX, and call in information. There are science reports, arts and culture programs, popular music, and features concerning events in Holland and Europe.

"Happy Station," Radio Nederlands' Sunday evening family show, has always brought a positive response from listeners everywhere. Happy Station host Tom Meyer reads listeners' letters and birthday greetings and plays music requests. In North America, you can hear Radio Nederlands evenings on 6.020, 6.165, 9.590, 11.705, and 11.835 MHz.

Radio New Zealand International

If you'd like to hear how things are going down in the land of the kiwis, just tune your shortwave radio to Radio New Zealand International at 9.700, 15.120, or 17.770 MHz. They'll read you the latest news, tell you the weather forecast for both North and South Island, and let you listen in on the most exciting rugby games you can hear anywhere.

But that's only the beginning. Because Radio New Zealand is currently relaying their home service over shortwave as a result of budget constraints, you'll get to hear the same thing as the locals when you tune in. Even though the vast majority of Radio New Zealand's programs are in English, you'll have no trouble recognizing that you're listening to a foreign country. For one thing, all the announcers have what sounds like a British accent. And they have quite a few expressions that are unique to their part of the world.

To most of us, a kiwi is a small, green, fuzzy fruit. But to a New Zealander, it can also mean a small, flightless bird about the size of a large chicken—or it can be a nickname for people from their own country, as in "The Kiwi Music Show," where only songs by New Zealand artists are played. Because New Zealand is in the Southern Hemisphere, the seasons are opposite to those in the Northern Hemisphere. So while people in North America, Europe, and Asia are shoveling snow, New Zealanders are enjoying summer. And instead of saying "we are expecting heavy rain" as an American or Canadian forecaster might, their meteorologist will tell you that "falls of rain are expected."

Because Radio New Zealand International is a relay of home service programming, you'll hear public service announcements, street closings, and dozens of other items that wouldn't be included in programs prepared for an international audience. About 8 percent of the New Zealand population is Maori. Their Polynesian ancestors crossed the Pacific in small boats like canoes and arrived in what is now known as New Zealand centuries before the Europeans. Even today, every Maori tribe is named for the canoe that brought their people to the islands.

Everyday, a segment of Radio New Zealand International's programming is devoted to Maori news, culture, and music. Young people in New Zealand and abroad can learn a few words of the Maori language if they tune into "Ears," a program of stories, music, and fun especially for children.

In addition to "The Kiwi Music Show," Radio New Zealand International broadcasts almost any type of melody you want to hear. They play classical music, songs from the big band era, Maori tunes, and plenty of modern pop and rock. Figure 4-6 is a QSL card from Radio New Zealand International.

Radio Czechoslovakia

One of the strongest signals coming out of Eastern Europe belongs to Radio Czechoslovakia. They broadcast half-hour programs in English to North America several times during the evening on 5.930, 7.345, and 11.990 MHz. Figure 4-7 is a Radio Prague International (now: Radio Czechoslovakia) QSL card.

Radio New Zealand International

Thank you for your report on our transmission

on 5 Oct. 1991 on a frequency of 9700

kHz which we are pleased to verify.

Our 100 kw transmitter is located

at Rangataiki near Taupo.

'Calling Japan' pilot series

GREETINGS FROM NEW ZEALAND

P.3060

4-6 This Kiwi welcomes you to tune in Radio New Zealand International!

Every program begins with news from Czechoslovakia and from around the world. Then you get to hear a review of the Czechoslovak press. After that, programming varies from day to day. You can hear features about life in Czechoslovakia, economic and political developments, nature and the environment, music (both traditional and modern), tourism, listeners' letters, and DX tips.

4-7 Card from Radio Prague International (Radio Czechoslovakia).

In 1989, when the brave citizens of Czechoslovakia filled the streets to protest communist rule, the networks were there to report on the happenings. They were there when President Havel was installed in office. But after the initial excitement died down, the North American media seemed to forget that Czechoslovakia even existed.

Changing from one form of government to another is never easy, especially after decades of damage caused by communist suppression. The problems that need to be solved are endless. But if you have a shortwave radio, you can tune in every night and find out how things are going through Czechoslovakia's own media outlet, Radio Czechoslovakia.

Vasclav Havel had never sought political power. But to the Czechoslovak public, he was the only man for the new job of president. He became a playwright in the 1960s, and used the theater as a means of ridiculing the Soviet regime. In 1968 when tanks from the USSR rolled into Czechoslovakia, he bravely took to the air on an unlicensed radio station to condemn the invasion. The Czech government responded by banning Havel's plays and ordering that all his books be removed from store shelves and libraries. He later served time in prison for his work in the Charter 77 human rights organization to push for basic liberties such as freedom of religious beliefs, freedom of expression, and freedom from government mail opening, phone taps, and house searches.

From the day he took office, President Havel warned the citizens of Czechoslovakia that they had an uphill battle ahead of them. The Soviet Union and the government it had imposed had done untold damage to the economy, environment, and national spirit.

Industries and businesses that had been taken over by the state are now being returned to private ownership. It takes time and endless parliamentary discussions to decide exactly how this can be accomplished most fairly. Further, many factories need to be renovated before they can turn out the types of products that will sell well on the foreign market.

There's a lot of work to be done. But it appears that these brave people, who dared to stand up against communist oppression, are up to the challenge. If you listen to Radio Czechoslovakia, you can hear it happen.

The Voice of America

The Voice of America is the United States Government's shortwave outlet to the world, and has been for the past 50 years. While VOA transmissions are intended primarily for a foreign audience, they have something to offer to American listeners as well.

You'll hear about events in the United States that our newspapers and TV networks miss. You'll hear about breakthroughs and inventions in the fields of science, medicine, and technology. You'll hear about fascinating people and places in our own country. And you'll hear how our government is presenting the United States to people in foreign lands. Figure 4-8 shows the VOA on election night, keeping the world informed as American voters select their new leaders.

4-8 The VOA, reporting on election night. VOA photo.

The Voice of America does everything it can to make its programs appeal to their intended audience. This approach includes playing popular local music, focusing on news events of the listening region, and hiring announcers from the part of the world they want to reach. Programs are produced in English, slow English, and dozens of foreign languages.

VOA programs are transmitted to satellites, then beamed to relay stations around the world. Their various overseas services can be heard on dozens of frequencies.

WWCR—Nashville, Tennessee

WWCR originally went on the air in the mid-1980s as a new shortwave outlet for Christian broadcasters. Much of their airtime is still taken up by religious programs . . . but that's only *part* of the story! They are now known as the worldwide voice of American opinion, airing programs with views from the political right, left, and everything in between. They're the only station in the United States where listeners can pick up the phone, call their favorite talk show, and be heard by shortwave listeners all over the world.

Sun Radio Network was one of the first groups to take advantage of WWCR's open programming policy, airing two national talk shows: Chuck Harder's "For the People" and Tom Valentine's "Radio Free America." And more recently, WWCR became a safe harbor for former pirate radio operators who have decided to go legal! Radio Newyork International, WWCR's first pirate show, came aboard on September 16, 1990.

RNI had its beginnings in 1987. When Allan Weiner and Pirate Joe found out that buying a station in (or anywhere near) the New York City area would cost far more than they could afford, they decided to put together an offshore station instead. With the help of some friends in New York's free radio community, they purchased a ship, named it the M/V Sara, and fitted it with transmitters and antennas for the AM, FM, and shortwave bands.

Offshore broadcasting had been going on in Europe for decades. The famous boat station, Radio Caroline, had been broadcasting from a ship anchored in the North Sea, off the coast of England, since the mid sixties. So it seemed logical that the same thing could be done off the coast of the United States.

Unfortunately, things didn't turn out that way. Even though the M/V Sara was clearly anchored in international waters and the station's transmitters were operated by licensed radio engineers, the FCC made a decision to close them down.

Only a few days after RNI's initial broadcast, the FCC moved in. They boarded the ship with Federal marshalls and Coast Guard personnel, had the crew arrested and removed from the boat—then proceeded to destroy much of the station's equipment.

Of course, the story doesn't end there. The RNI crew *did* find a way to get back on the air. You can hear them every Sunday night on 7.435 MHz, broadcasting over WWCR's 100,000-watt transmitter. They play music, take calls from listeners, do comedy skits, comment on important issues of the day, and read listeners' mail. Their programs are financed by selling ad time, by listener contributions, and by RNI T-shirt and video sales (Fig. 4-9). Tune them in from 10:00 P.M. to 1:00 A.M. Eastern Time—0300–0600 UTC. Other WWCR frequencies are: 12.160, 15.690, and 17.525 MHz.

4-9 RNI's Allan Weiner, Pirate Joe, and Johnny Lightning broadcasting over WWCR.

5

And now for the details

NOW THAT YOU HAVE A BASIC IDEA OF WHAT SHORTWAVE LISTENING IS ALL about, it's time to find out more about your new hobby.

Buying a shortwave radio

When you're ready to buy your first shortwave radio, the choices can be confusing. There are multiband radios with bands marked SW 1 and SW 2, inexpensive pocket-sized shortwave radios you can order for as little as $30, midpriced portable receivers priced in the mid-hundreds, and for the serious listener, there are top-of-the-line tabletop communications receivers with prices that can run into the thousands.

So which radio should *you* select?

Radios that try to cover everything

If you want to do more than hear a few high-powered international stations, cross off your list any multiband radio with bands marked SW 1 and SW 2. They try to cover everything: AM, FM, police, CB, weather, aircraft, TV, shortwave—you name it. And they usually do a very poor job of it all.

Some people buy multiband radios and use them for years, tuned to bands marked "police," "TV," or other words they can understand before they realize (often by accident) that "SW" means that you can bring in foreign countries! This type of radio can give you a taste of international listening, but that's about all. It has very poor selectivity and brings one station in on top of the other when band conditions are crowded.

Inexpensive pocket portables

The next class of receivers, name-brand pocket-size radios priced between $30 and $150, offer somewhat better reception. They can be a good "starter" radio for people who want to find out how much they like shortwave listening without having to invest a lot of money. You can hear stations from all around the world on them and have a lot of fun while you're at it.

And radios in this price range make a wonderful gift for a friend or relative you'd like to introduce to the hobby. The Grundig Traveler II (Fig. 5-1) has five shortwave bands and comes with a world time clock, a shortwave listening guide, and a case to store it all in whenever you're not using it.

5-1 Grundig Traveler II. Grundig.

The Sangean ATS 800 (Fig. 5-2) is one of the better-rated sets in the $120 to $150 price range. It has digital readout, 20 memories, and comes with an informative booklet on shortwave listening.

Medium-priced portables

To really upgrade your shortwave listening from the inexpensive portable level, be prepared to spend at least $200 for your radio. Sets in this price range are, in most cases, still small enough to fit into the palm of your hand, but they include some very important features less expensive radios neglect.

First, most midpriced portables have an SSB/BFO knob or switch, which enables you to hear ham radio operators, utility stations, and shortwave broadcasters who operate in single sideband. Sensitivity (the ability to pull in more stations) and selectivity (the ability to separate nearby signals so a strong station won't block out a weaker one on a nearby frequency) are better than what you would find on lower-priced sets.

5-2 Sangean ATS 800.

Nearly all radios in the medium price range have digital frequency readout, so you'll know exactly which frequency you're tuned to. Many have push-button frequency entry and memory banks (similar to scanners), making it easy to program your favorite stations' frequencies and hear them at the touch of a button.

Radios in this class have extendable whip antennas, so you can use them as they come out of the package. And if you decide to install an outdoor antenna later on, these receivers are designed well enough to handle the increase in signal strength without overloading.

The Sangean ATS 818CS (Fig. 5-3) does everything you'd expect of a medium-priced portable. It has digital tuning, a 45-frequency memory bank, a BFO control for single sideband and code reception, and as an added bonus, a programmable cassette player/recorder so you can save your favorite programs and share the sounds of international listening with your friends.

Tabletop communications receivers

After you've had a chance to familiarize yourself with the shortwave bands, you might decide to invest in a communications receiver. Prices start at several hundred dollars and extend up into the thousands. Communications receivers rarely come with extendable "whip" antennas, because manufacturers assume that anyone who is willing to invest that much money in a radio will want to install an outdoor antenna as well.

5-3 The Sangean ATS 818CS allows you to record and play cassette tapes.

With their elaborate bandwidth filters, scanning modes, and other high-tech features, communications receivers can reel in stations that lower-priced radios miss. But there is a downside. Communications receivers are much more sophisticated than portables, and it might take quite a while for a newcomer to the hobby to learn how to operate their numerous controls and functions.

But if you have the interest, some experience with shortwave listening, and the money, a communications receiver is one of the best investments you can make in your radio listening hobby. The Drake RF8, shown in Fig. 5-4, has 100 memory channels for you to store your favorite frequencies on, and a log in the back of the instruction book so you can record the station you've programmed into each one. The Icom IC-R71A shown in Fig. 5-5 is another excellent communications receiver.

A shortwave for your car

If you want to hear the world while you're on the road, the AM/FM/SW Philips 777 (Fig. 5-6) compact car stereo makes it possible. The only mobile shortwave radio sold in America, the Philips 777, covers shortwave frequencies from 3.170 to 21.910 MHz. It has digital tuning, a 20-station memory, and contains a cassette player/recorder so you can save all the great programs you hear.

5-4 The Drake RF8, a top-of-the-line communications receiver.

5-5 An Icom IC-R71A brings you the world. Icom photo.

Where to go for advice

Once you've decided how much you want to spend on your first shortwave radio, you still have quite a number of models to choose from. The *World Radio—TV Handbook* and *Passport To World Band Radio*—as well as *Monitoring Times* and *Popular Communications*—review new receivers as they come on the market.

Shortwave radio clubs and local amateur radio groups are another good place to go for advice. They are familiar with many of the models available from major radio manufacturers. And if you're lucky, you might come across a member wanting to upgrade, and eager to sell you the shortwave radio he's currently using, at a very reasonable price!

5-6 You can hear shortwave in your car with the Philips DC 777.

Trying out a shortwave radio

Before you buy a shortwave radio, it's a good idea to find out if you can take it back for a refund if it doesn't perform satisfactorily. If you try one out in an electronics or department store, you aren't likely to hear it at its best. Most store buildings are made of heavy materials that are difficult for radio waves to penetrate. If fluorescent lights are in use, they cause so much noise that you can't hear any stations, even if they are able to pass through the walls. And compared to what you'll hear in the evening when most foreign broadcasters aim their signal at North America, very little programming in English comes to our part of the world during the day.

NOTE: The Voice of America, Radio Canada International, The British Broadcasting Corporation, WWCR in Nashville, Tennessee, The Christian Science Monitor, Family Radio, HCJB Ecuador, and Radio Moscow should be heard broadcasting in English somewhere on the band at any time, day or night.

Understanding radio language

Long-distance radio listeners have a vocabulary of their own, much of it borrowed from early ham radio operators who used this form of radio shorthand to save time when sending messages in Morse code. If you enjoy listening to faraway stations, you are a DXer. DX means distance. But hearing a low-powered station at a moderate distance can be as much of a DX as hearing a higher-powered station broadcasting from the other side of the world.

Abbreviations and Q signals

SWL=Shortwave listener, anyone that listens to a shortwave radio. But it can also mean a person who prefers to hear major shortwave stations for the program content instead of scanning the bands for more difficult DX catches. A person known as a

DXer would be more interested in listening for low-power stations—weak, barely audible signals from domestic stations thousands of miles away. Most people involved in shortwave radio spend some time at both DX and SWL activities, then decide which one they like better.

If you're already involved in shortwave listening, you've doubtless heard Q signals mentioned. This form of radio shorthand was developed many years ago by ham radio operators so they wouldn't have to spell out frequently used messages letter by letter. Here are a few of the more common Q signals:

QRM = interference. If you want to tell a station that someone else is broadcasting on top of them, all you have to say is, "There was QRM from Radio ________ ."

QRN = static. Thunderstorms, power tools, and many other appliances can cause it.

QSB = fading. If the volume of the station you're listening to isn't consistent, there is QSB. It usually has to do with conditions in the ionosphere.

QSL = you already know what this one means: verification of reception. And it also means a nice card to display on your wall.

Collecting QSL cards

Part of the fun of being a shortwave listener is building a collection of attractive QSL cards from the stations you hear. But to receive QSL cards, you have to provide the station with some information that's very important to them: how well their signal is getting into your part of the world.

A well-written reception report gives the station information about:

- The frequency the station was using.
- Signal strength.
- Interference (if any) from other stations.
- Noise.
- Fading.
- Overall reception conditions.
- Date, time, and frequency of the program.

What time is it?

Shortwave stations get hundreds, if not thousands, of reception reports every week. These come from every part of the world and every time zone. So how do the broadcasters ever figure out when all those people were listening? By requesting that listeners use an internationally accepted worldwide standard time. It's known as Universal Time Coordinate, or UTC for short. Instead of the two daily 12-hour cycles used in the United States and some other parts of the world, the UTC system uses one 24-hour cycle. So there is no need to worry about A.M. or P.M.

Table 5-1 shows how to translate local times in North America into UTC. You can also get the correct UTC time by tuning into a standard time-and-frequency station, such as WWV in Ft. Collins, Colorado. You can find them on 5, 10, 15, and 20 MHz. They announce the UTC time every minute. WWV broadcasts come in on all four frequencies 24 hours a day. At night, try the lower frequencies of 5 or 10 MHz. During the day, standard time and frequency stations come in better on 15 and 20 MHz.

Canada, like many other countries around the world, has its own time-and-frequency station, CHU in Ottawa, Ontario. They broadcast on 3.330, 7.335, and 14.670 MHz.

Table 5-1. UTC time conversion for North American listeners.

UTC	EST	CST	MST	PST
0000	7:00 P.M	6:00 P.M.	5:00 P.M.	4:00 P.M.
0100	8:00 P.M.	7:00 P.M.	6:00 P.M.	5:00 P.M.
0200	9:00 P.M.	8:00 P.M.	7:00 P.M.	6:00 P.M.
0300	10:00 P.M	9:00 P.M.	8:00 P.M.	7:00 P.M.
0400	11:00 P.M.	10:00 P.M.	9:00 P.M.	8:00 P.M.
0500	Midnight	11:00 P.M.	10:00 P.M.	9:00 P.M.
0600	1:00 A.M.	Midnight	11:00 P.M.	10:00 P.M.
0700	2:00 A.M.	1:00 A.M.	Midnight	11:00 P.M.
0800	3:00 A.M.	2:00 A.M.	1:00 A.M.	Midnight
0900	4:00 A.M.	3:00 A.M.	2:00 A.M.	1:00 A.M.
1000	5:00 A.M.	4:00 A.M.	3:00 A.M.	2:00 A.M.
1100	6:00 A.M.	5:00 A.M.	4:00 A.M.	3:00 A.M.
1200	7:00 A.M.	6:00 A.M.	5:00 A.M.	4:00 A.M.
1300	8:00 A.M.	7:00 A.M.	6:00 A.M.	5:00 A.M.
1400	9:00 A.M.	8:00 A.M.	7:00 A.M.	6:00 A.M.
1500	10:00 A.M.	9:00 A.M.	8:00 A.M.	7:00 A.M.
1600	11:00 A.M.	10:00 A.M.	9:00 A.M.	8:00 A.M.
1700	Noon	11:00 A.M.	10:00 A.M.	9:00 A.M.
1800	1:00 P.M.	Noon	11:00 A.M.	10:00 A.M.
1900	2:00 P.M.	1:00 P.M.	Noon	11:00 A.M.
2000	3:00 P.M.	2:00 P.M.	1:00 P.M.	Noon
2100	4:00 P.M.	3:00 P.M.	2:00 P.M.	1:00 P.M.
2200	5:00 P.M.	4:00 P.M.	3:00 P.M.	2:00 P.M.
2300	6:00 P.M.	5:00 P.M.	4:00 P.M.	3:00 P.M.

For Daylight Savings Time, look ahead one hour on UTC column.

In the UTC time system, a new day begins at midnight. So you'll need to be careful to use the correct date on your report. For example, when there are still several hours left of Sunday in North and South America, it is already Monday in the time zone used as UTC, which crosses directly over Great Britain.

A signal-rating system

Shortwave hobbyists use the SINPO system, a form of radio shorthand, to evaluate reception conditions. It saves time for both listeners and station engineers, who all use the information contained in reception reports to get out the best signal possible.

"S," the first letter in the SINPO system, stands for signal strength. You rate the station's signal from 1 to 5, with 1 being barely audible and 5 being almost as loud and clear as a local broadcaster.

"I" means interference. If another station is causing interference to the one you're attempting to make out a report on, write down (if you can identify it) its name or station letters and which frequency it's broadcasting on. If there is heavy interference and it almost destroys the signal, put a 1 in the "I" column. If you can hear most of the program without any trouble but there is some interference, give it a 3 or 4. If there is no interference, rate it at 5.

The next column, "N," is for noise. In radio, noise is any sound on the band that isn't caused by man-made transmissions. Thunderstorms are one of the major causes of noise, especially during the spring and summer months. Sometimes they can cause static crashes in receivers hundreds of miles away.

But much of the noise you hear on shortwave radio is caused by man-made sources, such as fluorescent lights, motor-driven appliances, bad power poles, and the like. These devices plague both long-distance AM listeners and shortwave listeners, especially on the lower SW frequency bands.

If you are certain that a noise is caused by household appliances or other local sources, there is really no need to include it in your report, because the station has no way of controlling it. Stations can't do very much about thunderstorms either, except move to higher frequencies where atmospheric noise is less severe during the warmer months of the year, the time when storms are most likely to affect reception.

The letter "P" means propagation. If the radio wave propagation, or skip, is doing the job the station wants it to do, you'll hear the station just fine. But if it isn't, the signal will fade in and out as you listen. While "S" is used to indicate signal strength, "P" (or "F" on some report forms) refers to whether the signal fades in and out or stays at approximately the same volume.

"O" means overall performance. Usually, the rating for "O" is about the average of the other four items. It would certainly be no higher than the rating you gave the signal level, but it can be brought down to a much lower rating by just one signal-destroying item, such as interference or noise.

Writing letters that get read on the air

Mailbag shows are to a shortwave radio station what the "Letters to the Editor" page is to a newspaper. These shows are an open forum—an opportunity for you, the listener, to air your comments, views, and opinions to a worldwide audience.

Most shortwave stations receive hundreds, if not thousands, of letters from listeners around the world every week. Nearly all contain reception reports and requests for QSL cards, station souvenirs, and the like. A large number of them read something like this:

> Dear Radio XYZ,
>
> I like your station very much. Here is a reception report of your programs. Please send me a QSL card and whatever other items you have for your listeners.
>
> Thank you,

They are forwarded to the appropriate department and, in due time, processed.

But the letters that broadcasters most like to receive are the ones where listeners take the time to tell a little bit about themselves, their job or school, their radio receiving equipment, and their life in general. Letters like these let station staff members know who their listeners are. You're not just another name and address wanting a QSL card for his or her collection. You're a real human being who finds it worthwhile not only to take time out of your day to listen but, in the midst of your busy life, to sit down and compose a letter to the people that made the program possible.

When you tune in international shortwave mailbag shows, take note of what you hear. You'll soon see that the letters that *do* get read are more than just reception reports.

You don't have to turn out a literary masterpiece for your letter to get a slot on a mailbag show. You just have to *say something* that hundreds of other letters and QSL requests don't. (After all, no one would listen for long if the program host did nothing but read the names and addresses of people that requested QSL cards.)

If you can't think of anything to write, here are a few ideas to help get you started. As you go through the list, see what areas you think would make your letter most appealing, not only to station employees but to a worldwide listening audience.

- How long you have been a shortwave listener.
- How you got started.
- What type of radio and antenna you use.
- How old you are.
- What type of music you like.
- If you are a student, what is your favorite subject?
- Do you use shortwave radio to make school projects more interesting?
- What your occupation is.
- How you got interested in your career.
- If you have any children.
- If other family members listen to shortwave.
- What type of pets you own.
- What makes the area you live in special.
- Whether you belong to any interesting organizations.
- If you have any hobbies or collections that other people would like to hear about.
- What you liked best about the show you heard.
- Whether there was anything in the program that you didn't like.
- Whether you have a question that could be answered on the "mailbag" show or one of the station's other programs.
- Whether you have an idea for a new program.
- If you've traveled to the country you're writing to.
- Whether you think you might want to go there in the future.

Sending letters overseas

Once you have written your letter, the next job is to seal it in the envelope with your reception report and mail it! If you want to get your QSL card as soon as possible (and who doesn't?), be sure to send your letter by airmail. Surface mail can take months to arrive—and by then, your report will be of little value to the station (Fig. 5-7).

5-7 Send your reception reports by airmail, or they might end up traveling via banana boat!

Even when you send your reports by airmail, it usually takes at least a few months to hear from most stations. Economic problems have caused staff cutbacks at a number of shortwave outlets, making the reply time even longer. The vast majority of stations you'll hear as a new shortwave listener are financed by government or religious groups, and they're usually willing to pay the expense of mailing QSL cards and other souvenir items out to listeners.

When you're corresponding with a foreign country, the best way to pay for return postage is to use International Reply Coupons, also known as IRCs. You can buy them at your local post office. They serve as an international exchange medium and can be traded for domestic postage stamps throughout the world. In most countries, two or three IRCs are enough to send your QSL card via airmail.

Another important thing to remember when writing to a foreign station is to include your country in the return address. If you don't, the staff might have no idea where on earth to send their reply.

Corresponding with domestic stations

Once you'd had more experience with your radio, you'll start bringing in low-powered domestic stations, whose programs are only intended for listeners in their own region. Domestic shortwave stations often operate on a tight budget, and it's a good idea to enclose return postage when you write to them or you might never receive that sought-after QSL card.

Open my letter first!

While it doesn't guarantee a quicker response, making your envelope stand out from the rest can get your letter out of the mailbag faster once it reaches its destination. You can use colorful stamps and stickers, draw designs in ink or with colored pencils, purchase brightly colored envelopes, or even use fancy lettering as in Fig. 5-8. Anything you want to try is fine—just as long as the station address and return address are easy to read after you're finished.

5-8 A decorated envelope is sure to get attention.

Shortwave information at your fingertips

As a shortwave listener, you'll want to keep up-to-date on what's happening in the international broadcasting scene. You'll want to know when stations change frequencies, raise their power, or decide to use a new mailing address. And you'll certainly want to know when new stations are scheduled to take to the airwaves.

And with the recent surge in popularity of shortwave listening, you have dozens of sources of information to choose from. There are books, magazines, and clubs for every radio-related interest. Some are published by professional writers, while others are sold at cost by fellow hobbyists who want to share information with other listeners.

Books

Two of the most comprehensive frequency guides are *Passport to World Band Radio* and *The World Radio-TV Handbook*, both of which are published annually, and are available from nearly all book stores that carry radio publications.

If you're new to shortwave listening, *Passport to World Band Radio* is easier to use, because it focuses mostly on the stations whose programs are intended for an international audience. It gives general information on shortwave listening, reviews all the latest receivers, and contains a special list of the times and frequencies of stations broadcasting in English. It gives the addresses of major shortwave stations, as well.

The World Radio-TV Handbook contains the same type of information as *Passport to World Band Radio* and a whole lot more. In addition to covering the superpowers of the shortwave band, you get nearly 600 pages of information on radio, TV, and satellite broadcasting throughout the world.

If you're interested in DXing low-power outlets, you'll find plenty of frequency listings and addresses for domestic AM, FM, and shortwave radio services. You'll also read about satellite networks and TV DX opportunities. And if you want to write in for a QSL card, *The World Radio-TV Handbook* (*WRTVH*) can help you out. It gives the address of every station it lists.

Even if you don't have the type of equipment you need to pull in the more difficult DX catches, it's fun to browse through the *WRTVH* and see what's going on in broadcasting around our planet. For example, did you know that citizens of Great Britain can listen to Spectrum Radio, Downtown Radio, London Talkback Radio, Galaxy Radio, or Radio Harmony? And if you ever take a trip to New Zealand, the *WRTVH* will let you know where to tune in Access Radio, Radio Northland, Apple Radio, and Radio Bay of Plenty.

In 1992, Grove Enterprises came out with an annual *Guide to Shortwave Programs*. It lists 9,000 English language programs from more than 60 stations. Programs are listed by the day and time they can be heard.

Magazines

Every month, *Popular Communications* (Fig. 5-9) and *Monitoring Times* are filled with listening tips, station profiles, and updated frequency information. They cover shortwave, utility stations, pirates, AM, satellites, ham radio, scanning, and a number of additional radio-related topics. They review new products and publish dozens of advertisements that let you know where you can buy radio receivers, antennas, books, and just about anything else you need to get the most enjoyment out of your DXing hobby. Both magazines supply you with a monthly listing of shortwave stations broadcasting in English to North America.

- *Popular Communications*
  76 N. Broadway
  Hicksville, NY 11801
- *Monitoring Times*
  PO Box 98
  Brasstown, NC
  28902-0098

Tom Kneitel, editor of *Popular Communications*, and Bob Grove (Fig. 5-10), editor of *Monitoring Times*, have written a number of excellent books on radio hobbies and activities.

DX clubs

Whether you're into shortwave, scanning, AM DXing, or even pirate radio, there's a DX club that caters to your special interest. Most clubs release a monthly bulletin that's written, edited, and published by radio hobbyists who want to share the latest information on frequency changes, QSL policies, and programming with other listeners.

5-9 *Popular Communications* keeps you up-to-date on radio listening activities.

5-10 Bob Grove, editor of *Monitoring Times* magazine.

While commercial radio magazines rely on some input from their readers, the bulk of columns and articles are regular features, submitted by writers who are well known in their field of expertise. But shortwave club bulletins and newsletters are nonprofit publications—written, published, and mailed out by fellow DXers who want to share their knowledge and make friends with other hobbyists throughout the country.

If you join a DX club, you can receive current information on your favorite radio activities and bands. Most clubs come out with at least one bulletin a month. (Some clubs also run computer bulletin boards of up-to-the-minute DX information.) With the close cooperation of members and volunteer editors, they are able to print and distribute the latest DX news much quicker than most commercial publications can. And not only that. If you take the time to mail your loggings in to the editor, you'll have the thrill of seeing your name and "catches" in next month's issue! You'll find a worldwide listing of DX clubs in Appendix B.

Shortwave information on your radio

Your shortwave radio is one of the most up-to-the-minute sources of DX information around. Nearly every major station has its own DX program, supplied with the latest frequencies, time changes, and general radio news—by a worldwide audience who write, call, and fax in their latest listening tips. See Table 5-2.

Table 5-2. DX programs that are broadcast in English.

Country	Program	Day (VTC)	Time*
R. Australia	"Communicator"	Sunday	1430
		Monday	0730
		Friday	0430
			1030
R. Austria		Sunday	1130
			1530
BBC World Service	"Waveguide"	Saturday	0905
		Monday	0530
		Thursday	0130
BRT Belgium		Sunday	0035
		Monday	1437
R. Budapest, Hungary		Sunday	0235
		Wednesday	0235
HCJB, Quito, Ecuador	"DX Party Line"	Sunday	0039
			0239
			0509
R. Nederlands	"Media Metwork"	Thursday	1154
			1454
			1654
		Friday	0054
			0354

Table 5-2. Continued.

Country	Program	Day (VTC)	Time*
R. Prague International (Czechoslovakia)		Friday	0012
			0116
			0316
R. Sofia, Bulgaria		Monday	0030
			0545
			1930
		Thursday	0030
			0545
		Friday	1930
			2100
		Saturday	0030
R. Japan	"DX Corner"	Sunday	0330
			1530
R. New Zealand	"Authur Cushion's DX Tips"	Every other Monday	0430
		Thursday	0830
		Friday	1930
R. Sweden	"Sweden Calling DXers" (biweekly)	Tuesday	1313
			1543
			2343
		Wednesday	0113
			0213
Swiss Radio International	"Swiss Shortwave Merry Go-Round"	Saturday	0648
			1118
			1318
			1348
			1548
		Sunday	0218
			0418
Trans-World Radio (Nederlands Antilles)	"Bonaire Wavelength"	Sunday	0300
		Saturday	1130
Voice of America	"Communications World"	Saturday	1210
			1710
			2110
		Sunday	1010
WWCR (USA)	"Signals"	Sunday	0435
	"Crossband"	Monday	0200
	"World of Radio"	Sunday	0405

*Times given in UTC (Universal Time Coordinate).

(Note: Check for these programs an hour earlier during daylight savings time.)

Hey, why can't I hear that?

No matter what source of frequency information you use, you're going to read (or hear about) a number of stations that you just can't seem to pull in. The reasons vary. For one thing, stations can, and often do, change their frequencies. Radio-magazine and DX-club editors try their best to make sure the information they publish is correct. But frequencies are sometimes out of date before they reach your mailbox.

Secondly, many of the people that provide frequency information to the publications you receive probably don't live in the same area you do. People living on the East Coast have a better chance of hearing stations broadcasting from Europe and Africa, while West Coast listeners have a better shot at bringing in stations from Asia and Australia.

And then of course, there's always the DX reporter that can afford a top-of-the-line receiver and has enough space on his property to put up the kind of antenna system most hobbyists can only dream about. He reels in DX catches faster than he can write them down in his logbook.

But no matter where you live or what kind of equipment you're using, there are plenty of DX targets out there on the bands. Reception conditions are always changing, and if you're alert about what to expect from your favorite frequencies, you'll know when you've found something out of the ordinary. Just keep trying, and sooner or later, you're sure to come up with some interesting DX catches!

Putting up an antenna

If you decide you want more of an antenna than the extendable "whip" that comes with your portable—or are investing in a communications receiver that has no antenna of its own—you have several options.

Your antenna choices

Anyone with the necessary space can erect a long-wire antenna. All you'll need is about 75 feet of copper wire, a few insulators, and some extra wire to make a "ground" connection from the "GND" terminal on the back of your receiver to a cold water pipe. Figure 5-11 shows how to do it.

If you live in an apartment and your landlord won't give you permission to put up an outdoor antenna, you can string wire around your room, connect it to the receiver, and see if reception improves.

Many shortwave mail-order companies sell antennas especially designed for the shortwave frequency bands. They also carry active indoor antennas that add amplification to incoming signals—and can significantly improve reception.

Most portable radios in the "inexpensive" category aren't designed well enough to keep powerful stations from splattering all over the band when you hook them up to an outdoor antenna. They usually don't even have an external antenna connection. But if you want to try anyway and see what happens, you can make the connection by wrapping the lead-in of the external antenna around the extendable whip antenna several times.

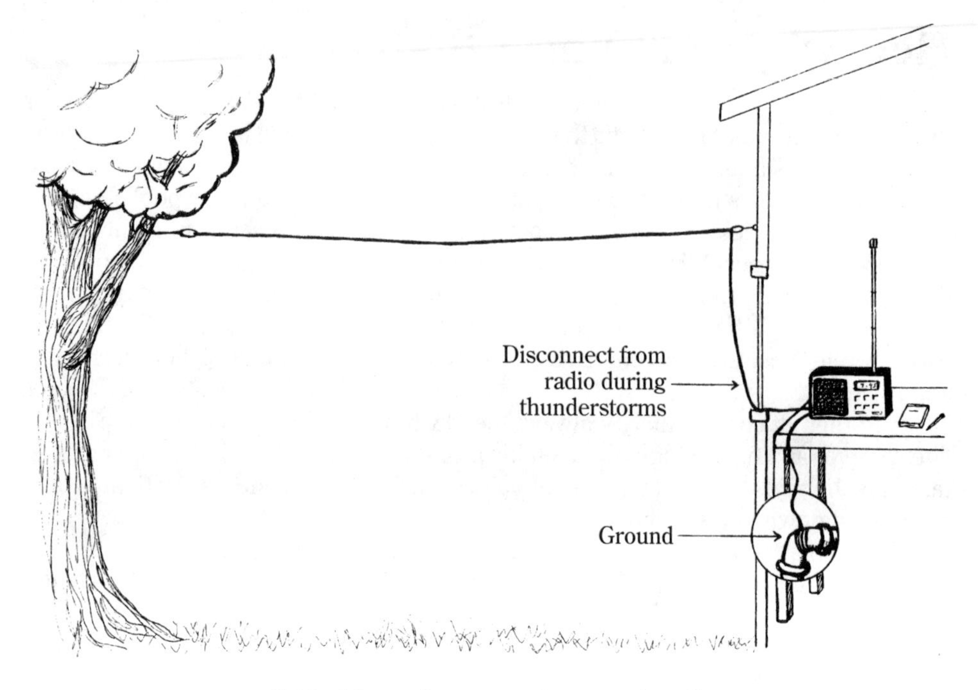

5-11 A long-wire antenna is easy to install.

Important things to remember

Whenever you erect an antenna outdoors, you should try to get it up as high as possible. This allows more signal to travel into the wire, and into your receiver. Also, it's important to keep your antenna as far away as possible from all obstacles, *especially* power lines!

6
Tuning in the action bands

DON'T YOU WONDER WHAT'S GOING ON WHEN A POLICE CAR, FIRE TRUCK, OR ambulance speeds down the street with its siren blasting? If you're tired of waiting for the evening news to find out what happened (and who it happened to), a scanner should definitely be part of your radio listening equipment.

A scanner keeps you informed

Scanners are programmable radios that bring in emergency communications and an assortment of other stations in the VHF-UHF frequency range. With a multiband "police radio," you are limited to hearing one station and one frequency, until you take the time to tune to another one.

But scanners have the advantage of checking out a number of frequencies in a very short time period. They "scan" the frequencies you program, stopping only when they locate a transmission in progress. You select the frequencies, punch them in, and the scanner will do the rest.

What you can hear

Scanners bring you police calls, fire departments in action, emergency rescues, utility company transmissions, ham radio operators talking to friends across town, taxicabs, ambulances, hospital helicopter services, business radio, railroad workers—even airline pilots, giving their current position, reporting wind conditions, or checking with the control tower to see if they have clearance for a landing.

Buying a scanner

Basically, scanners come in two types: portable and base. Portable scanners are battery-powered, hand-held units that often come with rechargeable battery packs. Base unit scanners sit on your desk and operate off standard household current. Both types come with their own antennas, although you can hook up an external antenna for better reception if you wish.

Figure 6-1 is a portable Realistic Pro-37 scanner. It scans at the rate of 25 frequencies per second and accepts up to 200 frequencies in its memory bank.

6-1 Realistic Pro-37 scanner.
Radio Shack.

How VHF/UHF signals travel

Signals transmitted on the VHF (Very High Frequency, 30 to 300 MHz) and UHF (Ultra High Frequency, 300 MHz to 3 GHz) bands travel in a line-of-sight path—meaning they aren't able to bend and follow the curve of the earth, as AM and lower-band shortwave frequencies can. You can extend your scanner's range by installing an outdoor antenna, but unless unusual "skip" conditions are present, your normal listening range will probably be restricted to about a 50-mile radius.

Still, there's a lot to hear!

Hearing skip on your scanner

Occasionally, conditions in the atmosphere make it possible for you to hear signals in the VHF/UHF range that are far beyond the normal line-of-sight range. This is especially true during years of high sunspot activity when the E-layer of the ionosphere occasionally becomes dense enough to bend radio waves on the VHF band back towards earth.

Freak weather conditions also cause VHF skip. Temperature inversions create what is known as a "tropo duct." Once VHF signals get caught between the layers of air, they can be transported to listeners hundreds of miles away.

VHF "skip" signals frustrate local police, fire, and other emergency service dispatchers, especially when the signals are strong enough to be confused with local calls. It's nearly impossible to identify the location of out-of-town emergency calls. They are short, to the point, and can fade out as fast as they appeared. But they can certainly make some very interesting listening.

Finding action frequencies

On the local level, many scanner dealers can supply you with a list of police, fire, ambulance, and other VHF/UHF transmissions in your area. When you listen to your scanner, remember that transmissions aren't continuous, as they are on AM, FM, and shortwave. In fact, most VHF/UHF scanner stations (especially emergency services, such as fire and police) endeavor to keep transmissions as short as possible. Then the frequency is open for the next call.

Books, magazines, and clubs

If you hope to catch and identify some skip transmissions, or just want to see if you're missing out on some good frequencies in your local area, the following books can be helpful:

- *Police Call Radio Guide* - nine regional volumes.
- *Betty Bearcat National Police Frequency Directory.*
- *Top Secret Registry of U.S. Government Frequencies (25 to 470 MHz)*, by Tom Kneitel.
- *The Citizen's Guide to Scanning*, by Bob Kay.

The National Scanning Report, published by the Bearcat Radio Club, is an excellent magazine for scanner enthusiasts of all levels of experience. It's filled with true stories of emergency scanner frequency operations across America. It also gives listening tips and reviews of the latest scanning products.

You might also want to join a local scanner club. Appendix B lists several regional radio clubs that focus mainly on VHF/UHF activities.

10-Codes

Most police, fire, ambulance, and other emergency service departments use a 10-Code to save time when relaying information. Table 6-1 shows the official 10-Code list recommended by The Association of Police Communications Officers. But the APCO code list isn't universally accepted. Many areas still use their own 10-Code. Your local scanner dealer is the best place to obtain a copy of the codes your law enforcement and emergency agencies use.

Scanning the skies above us

From the time they're preparing for takeoff until the time they come in for a landing, jet aircraft, small planes, and helicopters keep in contact with tower and ground stations. Air traffic on the VHF band can be heard between 118 and 136 MHz—just above the FM band on the radio frequency spectrum.

Did you ever wonder why you can hear jets so far away? Well, as you already know, the higher you can get your receiver's antenna up, the farther away the stations that you can hear. This is especially true when you're dealing with high frequencies that only travel in straight lines. The same goes for the transmission tower—the higher up their antenna is, the farther the signal goes out. So if a broadcaster put up an antenna several miles high, you can easily hear his transmissions hundreds of miles away.

That's exactly what's happening when you tune in a jet aircraft on the VHF band! A signal transmitted from such high altitudes can cover an amazingly wide territory. That explains why you can hear the ground control (airport) station only within several miles of the tower, but broadcasts from jet planes in-flight can be heard several states away. And it also explains why you can hear jet planes much farther away than you can hear small, lower-flying private planes (Fig. 6-2). Even if you only have an inexpensive multiband radio that covers the VHF aviation band, you can expect to hear a good number of aircraft transmissions.

What you'll hear on aero frequencies

Although takeoffs and landings (see Fig. 6-3) are the most critical times for radio communication—especially at busy airports!—in-flight communication is also important. It lets ground stations know how planes are progressing on their journey.

Table 6-1. Association of Police Communications Officers 10-Code.

10-1	Signal weak
10-2	Signal good
10-3	Stop Transmitting
10-4	Affirmative (OK)
10-5	Relay (to)
10-6	Busy
10-7	Out of Service
10-8	In Service
10-9	Say Again
10-10	Negative
10-11	_______ on duty
10-12	Stand by (stop)
10-13	Existing conditions
10-14	Message/Information
10-15	Message delivered
10-16	Reply to message
10-17	Enroute
10-18	Urgent
10-19	(In) contact
10-20	Location
10-21	Call _________ by phone
10-22	Disregard
10-23	Arrived on scene
10-24	Assignment completed
10-25	Report to (meet)
10-26	Estimated arrival time
10-27	License/permit information
10-28	Ownership information
10-29	Records check
10-30	Danger/caution
10-31	Pick up
10-32	______ units needs (specify)
10-33	Help me quick
10-34	Time

Commercial jetliners fly along "skyways" in the air. Although there are no road signs, high-tech instruments help pilots keep their planes at the correct position on these imaginary highways. As they move on towards their destination, pilots keep ground control stations updated as to their altitude and position on the flight path. They also report unstable wind conditions that other pilots would want to avoid.

Pilots give this information in a form of aviation shorthand so the frequency will be clear for the next transmission. Most in-flight reports include:

- The airline name.
- The flight number.
- The flight level (altitude).
- Any unusual wind conditions.

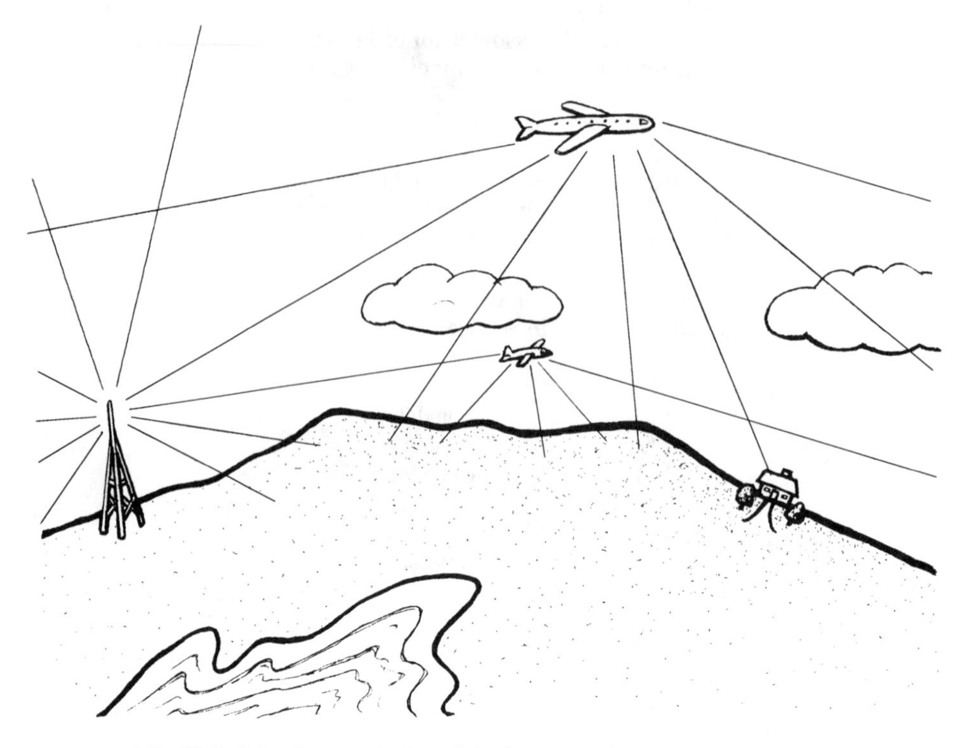

6-2 High-flying jets can be heard farther away than small private planes.

6-3 A Delta 727 jet preparing for takeoff. Jean Baker.

The airline name and flight number help ground control workers make sure they know which plane they are tracking.

The flight level, given in three numbers, such as 230, is shorthand for 23,000 feet. (To come up with the right altitude, all you have to do is add two zeros to the flight level number.) And strong or unusual wind conditions are always of interest to other pilots on the skyway who will want to try another flight level to give their passengers a smoother trip.

If pilots have an emergency situation, such as a sick passenger, they let ground control operators know so that a medical team can be on hand as soon as the jet makes its landing. Minor problems, for instance when a crew member needs to pick up a new uniform to replace a torn one, are also reported over the air.

Ground stations guide aircraft

Although you can only hear the pilot's side of most aero conversations, it helps to know a little about the ground stations they are in contact with. The main types are:

- Air route traffic control centers—they're in charge of directing commercial, private, and military aircraft throughout most of their trip. They identify themselves by the name of their city. Most ARTCCs work with aircraft flying over a several-state area. Often, you'll hear pilots refer to the name of the ARTCC they are calling.
- Aeronautical en-route ground stations—used for commercial pilots to communicate with their companies during flights.
- Approach control—communicates with pilots within 40–50 miles of the airport where they intend to make a landing. With the help of radar screens, Approach Control handles flights until they arrive at the outer edge of the airport. Then, control is given over to the local tower control.
- Local tower control—issues landing instructions, advises pilots of weather conditions and other traffic in the vicinity, and gives final clearance to land.
- Ground control—when an aircraft is on the ground (before and after a flight), it comes under the jurisdiction of ground control. This station directs traffic in the gates, taxiways, and runways. Figure 6-4 shows ground traffic controllers at work.

Finding aero frequencies

Table 6-2 gives a rundown on aero frequency allocations on the VHF band. Your local airport personnel will probably let you in on the frequencies they use for takeoff and landing communications. For more information on in-flight, tower, aeronautical telephone, military, emergency, and other special aviation frequencies, consult:

- *Air Scan—Guide to Aeronautical Communications*, by Tom Kneitel, K2AES.
- *Official Aeronautical Frequency Directory*, by Robert Coburn.
- *Directory of Military Aviation Communications* (*VHF/UHF*), by Jack Sullivan (separate directories are printed for each region).

6-4 Air traffic controllers make sure that planes stay on the right track. Jean Baker.

Monitoring Times magazine also has an aero band column—"Plane Talk," by Jean Baker.

Out-of-this-world transmissions

While high-flying jets can transmit over several states, really far-out signals can be heard in even wider areas. In the past several years, many scanner enthusiasts throughout the world have had the opportunity to hear astronauts on space shuttle missions when they operated amateur radio equipment on 144.55 MHz, in the 2 meter ham band. Even though they are hundreds of miles above us, their voices come in loud and clear.

Table 6-2. Aero frequency allocations on the VHF band.

Frequency	Purpose
108–117.975 MHz	Voice and instrument aeronautical navigation.
118–121.4	Airport air traffic control towers and air route traffic control centers.
121.5	International VHF emergency frequency, used for both voice and emergency locator transmitter broadcasts.
121.6–121.95	Ground control—runway and other ground traffic.
121.975–122.675	Flight service stations for private aircraft. These furnish information on airport conditions, process flight plans, etc.

Table 6-2. Continued.

Frequency	Purpose
122	Flight watch weather stations.
122.1	Flight service station call frequency for private and commercial aircraft.
122.2–122.6	Flight service stations to private aircraft.
122.4–122.7	Private aircraft to air traffic control towers at large airports.
122.725	Private airfields not open to the public.
122.750	Air-to-air communication.
122.8–122.95	UniCom aeronautical advisory stations (small fields). Gives information on weather, runway conditions, fuel availability, and so on.
122.9	MultiCom (temporary communication for small fields without UniCom) and air-to-air.
123.025	Helicopter air-to-air.
123.5–123.075	Heliports.
123.1	Civil Air Patrol and Coast Guard search and rescue.
123.450	Commercial airline air-to-air communications.
123.5–123.575	Flight test and miscellaneous.
123.6–123.65	Miscellaneous communications.
123.675–128.8	Air traffic control.
128.825–132.0	Aeronautical en route/company communications.
132.025–136.975	Air traffic control.

7
Getting into ham radio

IF YOU THINK YOU MIGHT LIKE TO DO SOME TRANSMITTING YOURSELF ,WHY NOT get into ham radio? Once you become a licensed ham radio operator, you have a world of communications options. You can contact people around the world on the high frequency (shortwave) ham bands. Or if you prefer, you can talk to people in your own local area on VHF/UHF frequencies.

While most hams still make most of their contacts with voice or Morse code, some amateur radio operators incorporate the latest technological developments into their stations. They use satellites and computers and other high-tech devices to get their messages out across the airwaves.

At one time, learning CW (Morse code) was the #1 stumbling block that kept many people from applying for a license. Now, with a No-Code Technician Class License available, newcomers to ham radio can gain access to the VHF and UHF bands without memorizing the "dit-dah" system.

Ham radio has something for everyone

Some people think all hams are electrical engineers, talking shop over the airwaves and experimenting with radio equipment in their spare time. While some amateur radio operators do earn their living in electronics, you're just as likely to meet a teacher, farmer, businessman, radio announcer, housewife, railroad worker, or student when you go on the air. You might even meet a senator (Fig. 7-1) or an astronaut (Fig. 7-2) if you're lucky! Amateur radio is a hobby with so many facets that it attracts a wider range of people than you might imagine.

Hams on the 2 meter band have round table discussions with friends around town. Operators on the shortwave frequency (3–30 MHz) ham bands chat with people across the country and, when the skip is right, around the world. Some operators

7-1 Barry Goldwater, K7UGA, a well-known ham radio enthusiast.

meet on a prearranged frequency every night to relay messages across the country from one operator to another.

If you're into experimenting with electronic gadgets, you can build or modify your equipment and then test it on the air with your friends. Or, if you'd rather have an outfit you can take out of the box, plug in, and use, there are plenty of ready-made

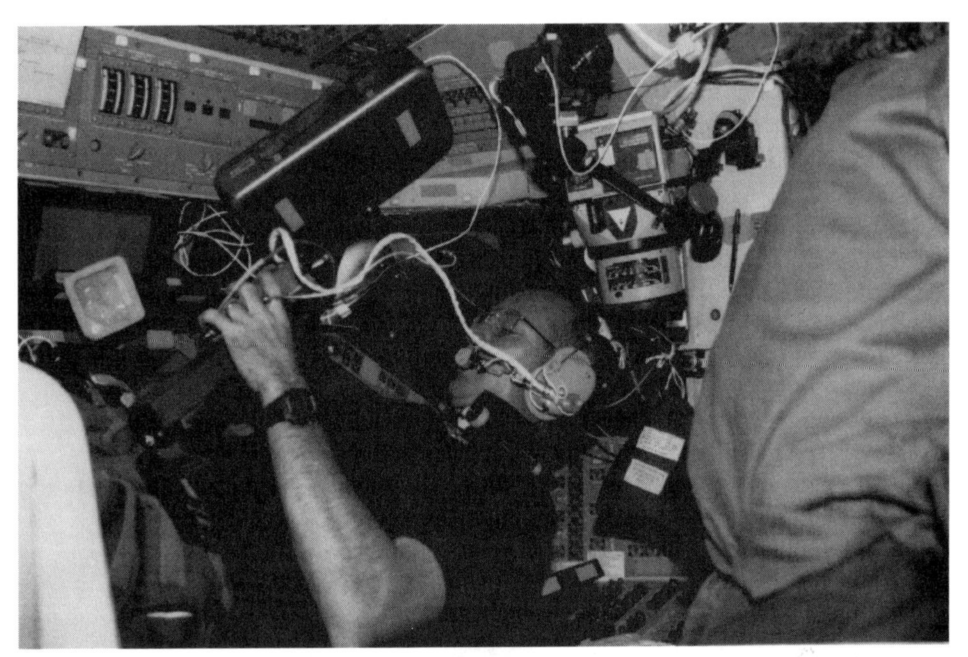

7-2 Astronaut Brian Duffy talks to hams on earth during a recent mission aboard Space Shuttle Atlantis. NASA photo.

transmitters, receivers, and transceivers (transmitters and receivers combined) on the market to select from.

The 2 meter band

The 2 meter band is the most popular meeting place for No-Code Technician licensees. And with a hand-held transceiver, such as the Realistic HTX-202 (Fig. 7-3) or the Icom IC-24AT (Fig. 7-4), you can get on the air without even having to worry about putting up an external antenna. If you have a scanner or multiband police radio, you can listen in on 2 meter ham activity between 144 and 148 MHz.

Repeaters get your message across

Since VHF/UHF signals only travel in a straight line, they sometimes need a little help getting to their intended destination. Repeaters are common on the 2 meter band. They're installed (usually by a ham radio club) in the highest location possible—on top of a hill or a tall building. They use one frequency to receive and another frequency to amplify and retransmit the signal (Fig. 7-5). When using a repeater, you can often contact people forty or fifty miles away.

Save on cellular phone expense

If you keep in contact with family and friends via cellular phone, ham radio can save you money. Once you get your license and purchase a mobile or hand-held UHF/VHF unit, you can keep in touch without having to deal with cellular phone companies or

7-3 Radio Shack's Realistic HTX-202 gets you on 2 meters.

7-4 The Icom IC-24AT.
Icom photo.

pay their expensive fees. Since many repeaters have autopatch capability (meaning they can dial telephone numbers), you can use your transceiver to communicate with hams as well as non-hams.

Computers on ham radio

Packet radio communication on the 2 meter band allows you to use your computer to communicate with other ham/computer enthusiasts. Your computer is connected by way of a data controller box to your transceiver. Messages are composed on the key-

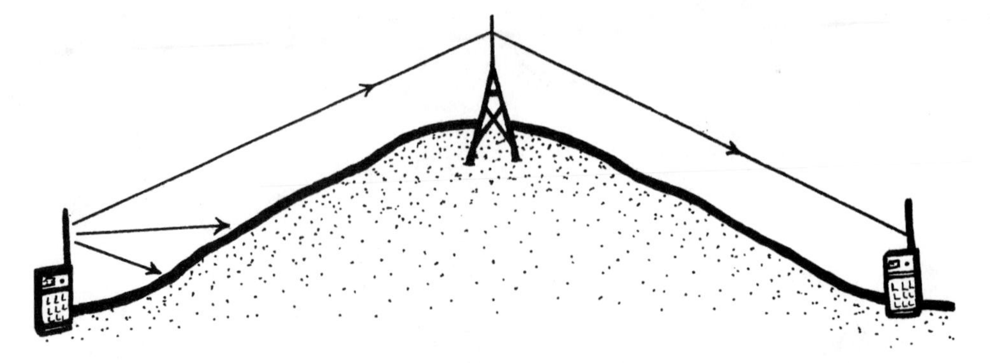

7-5 Repeaters amplify and retransmit signals on the 2 meter ham band.

board, then sent over the air in bursts of data called "packets," addressed to a specific station.

If an operator you want to contact is too far away to receive a direct transmission, packet data can be held and retransmitted by other hams through packet networks until the data reach their destination. Many repeaters are designed to receive, amplify, and retransmit packet radio messages. They also accept and hold messages until the operator they are addressed to checks in.

Hams come to the rescue

One purpose of the amateur radio service is to provide reliable communications in case of an emergency. Every year, ham radio operators across the country practice with emergency rescue teams so they'll be ready to go into action in case a real disaster strikes.

Hams on TV

In areas large enough to attract sufficient interest, hams set up TV stations on the VHF/UHF bands. They can transmit still or moving TV pictures to operators across town. Since hams aren't permitted to use the frequency bands allocated for commercial TV, you'll have to purchase special transmitting and receiving equipment if you want to get into this aspect of the hobby.

Getting your license

Of course, before you can go on the air, you'll have to pass an exam, and then wait several weeks for your license to arrive in the mail. To get ready to take the No-Code Technician exam, you have the option of attending a class or ordering the materials and studying by yourself (or with a friend) at home.

If you decide to attend a class but don't know where to find one, contact the American Radio Relay League or the National Amateur Radio Association (see Appendix B for addresses and phone numbers). They can provide you with a list of clubs and volunteer examiners in your area. If the individuals listed aren't teaching classes themselves, they can recommend to you another operator who is directly involved in helping people enter the hobby.

To pass the No-Code Technician test, you'll have to be able to answer 41 out of 55 questions correctly. The FCC test, administered and supervised by local Volunteer Examiner (VE) Coordinators, covers electronic principles, radio wave propagation, and FCC regulations.

Your VE will select the questions used in your test from a pool of 700. The questions are widely published, and available from a number of the sources listed below. Over 5,000 people a month in the United States are getting on the air the "No-Code Tech" way!

Ham radio license study guides

The materials listed below can help you pass the No-Code Technician test, whether you are taking classes or studying at home on your own:

Now You're Talking
No-Code Technician Class license study guide. Order this title from:
American Radio Relay League
225 Main Street
Newington, CT 06111

Morse Man Plus—code training software for IBM-compatible computers. You can get this from:
Renaissance Software & Development
Kilien Plaza
Box 604
Kilien, AL 35645
Phone: (205) 757–5928

No Code Ham License. Write to or phone:
Diamond Systems, Inc.
PO Box 48301
Niles, IL 60648
Phone: (312) 763–1722

National Amateur Radio Association Education Package (handbook with IBM or Macintosh software). Includes a complete list of volunteer examiners, the FCC rules and regulations booklet, and a sample copy of *The Amateur Radio Communicator* magazine. Write or call:
NARA
16541 Redmond Way
Suite 232
Redmond, WA 98052
Phone: (800) Got–2–HAM

No-Code Ham Radio (textbook and IBM software). Order from:
W5YI Group
Box 565101
Dallas, TX 75356
Phone: (800) 669–9594

Where you can transmit

Figure 7-6 shows amateur radio frequency allocations for all ham radio license classes in the United States. If you are going for a No-Code class license, you'll have access to all amateur frequencies above 30 MHz.

The most popular VHF/UHF bands for No-Code Tech licensees are 2 and 6 meters. Local radio clubs can tell you which frequencies are most active in your area.

Call letters for your station

When the FCC issues your ham radio license, they'll give you a set of call letters to use to identify your station when you're on the air. A ham radio call consists of several letters and a number.

The call letters that the FCC issues you depend on where you live, when you apply (license letters are issued in alphabetical order), and the class of license you apply for. The number in your call letters tells other operators the region you live in.

Figure 7-7 is a map of United States amateur call zones.

Your QSL card

If you've been writing in to the AM and shortwave broadcast stations you hear, you probably have a nice start on a QSL collection. Well, when you become a ham, the operators you contact will want a QSL from *you*. (This is especially true if you live in a "rare" state or county that has a low ham population.)

Once your license arrives and you've had a chance to get on the air and try out your equipment, your next order of business should be to get some QSL cards printed.

Several companies specialize in designing and printing ham radio QSL cards. They advertise in the major amateur radio magazines and will send out a catalog of their standard designs on request. But if you want to save money—and, at the same time, make sure your card is different from everyone else's—you can design your own QSL and have it printed at a local shop.

Designing a QSL

Of course, all cards need to have a place for the basic QSL/QSO information:

Call letters / Date / Time / Frequency / Mode

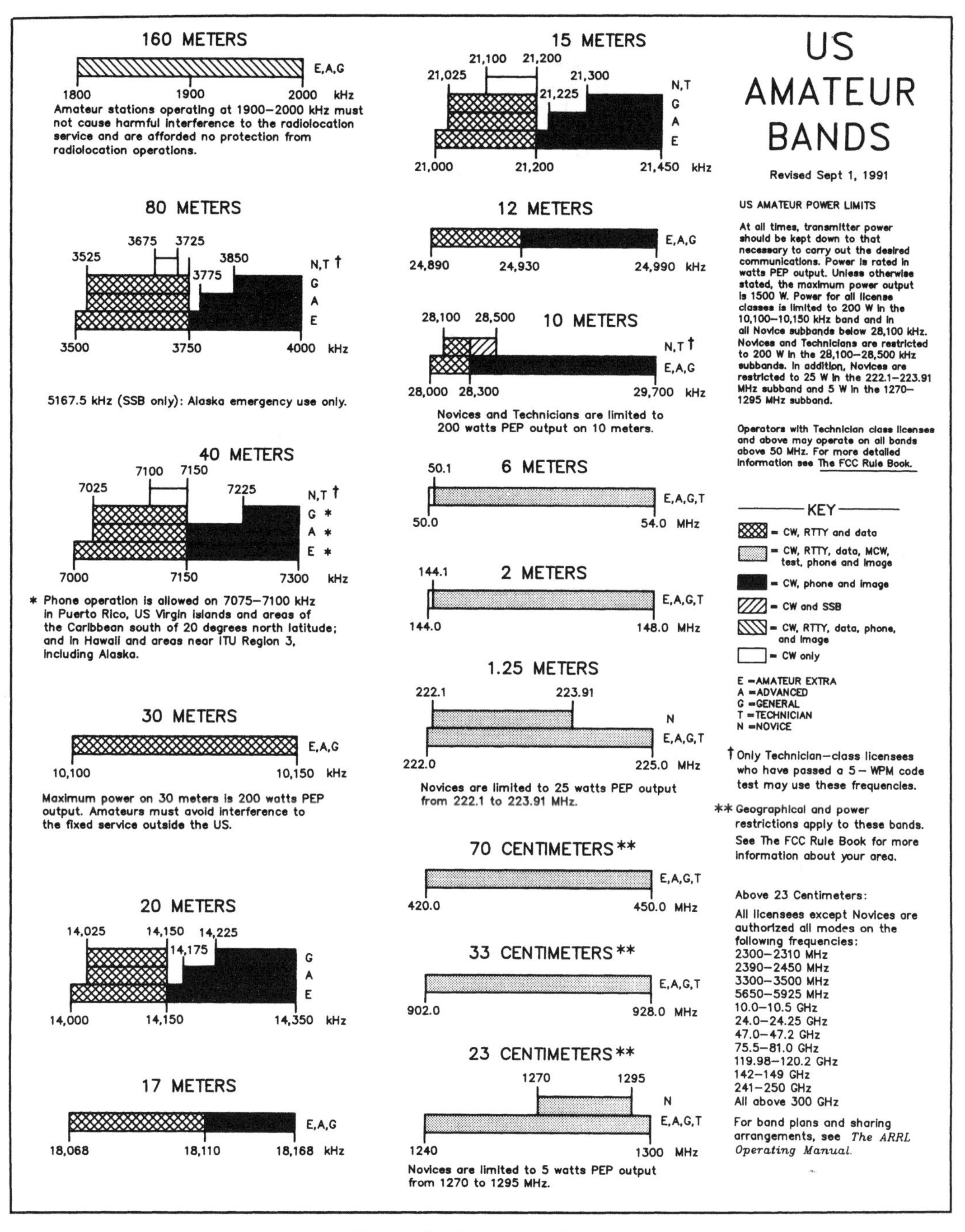

7-6 Ham radio frequency chart. ARRL.

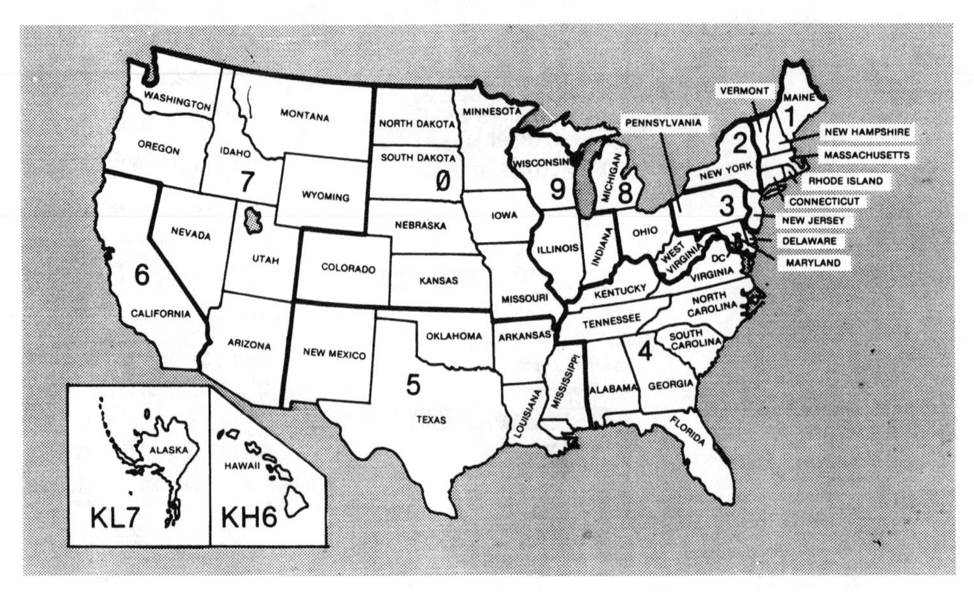

7-7 Map of U.S.A. call number zones. ARRL.

After that, it's all up to you! Just remember: the cards you send symbolize you and your station to everyone that requests your card to display on his or her radio room wall.

The first thing you'll want to do is decide what kind of design to use. Many hams select designs that have something to do with ham radio—antennas, globes, drawings of transmitters, radio waves, and so on. If your town is known for something special, you can incorporate that into the design, as W3FVF did (Fig. 7-8).

7-8 W3FVF wants you to know what his town is famous for!

If you are artistically inclined, you have the option of doing your own artwork. It's a good idea to use black drawing ink, as it shows up better than felt-tipped markers, ballpoint pens, or pencils when the cards are printed.

Or you can clip a black and white design from another source for your QSL card. Although it is much more expensive than using a standard QSL print shop design or designing your own card, some hams order individualized full-color QSL cards that feature photographs of themselves, their location, or whatever. If you want to put the QSO information (QSO = communication between two stations) on the decorated side of the card, you'll have to allow some room for it before you plan the rest of the design. And whichever design plan you choose, your call letters should be clearly visible on the front.

Figure 7-9 shows one way of designing a QSL. You can have your local print shop take care of the lettering. Or if you have a computer and printer, you can save on the typesetting costs by doing the whole job yourself with a graphics or a desktop publishing program.

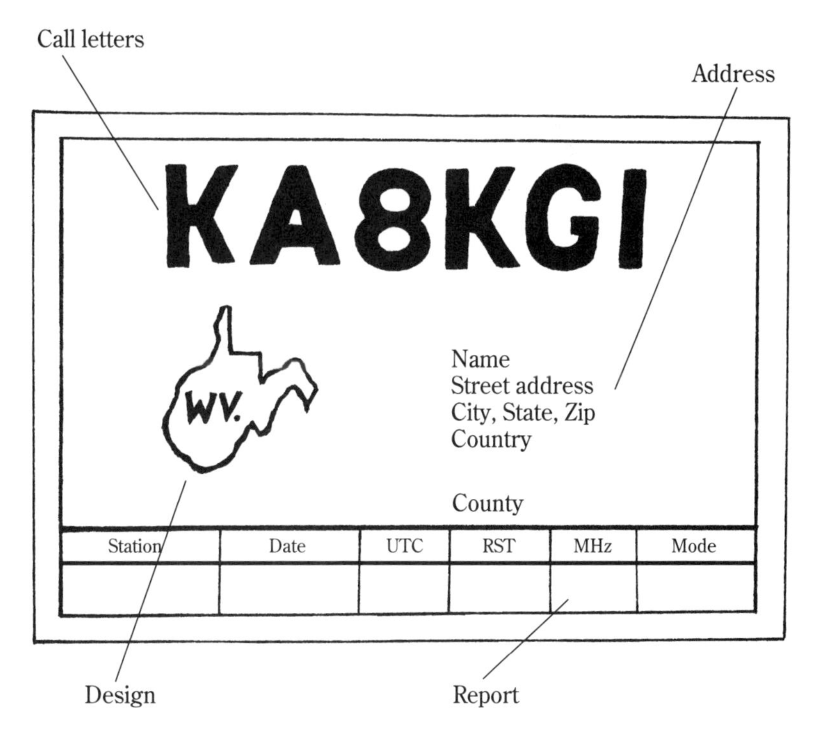

7-9 You can design your own QSL card—and save lots of money.

Higher licenses = more frequencies

After you've been on the air for a while as a No-Code Tech operator, you'll probably be anxious to get a higher class license with privileges on the shortwave "skip" bands. To do that, you'll have to learn Morse code and pass a written exam covering electronic principles, operating procedures, radio wave skip, and FCC regulations for using the bands you'll now have access to.

The easiest way to get on the shortwave (high-frequency) bands is to have a Novice License. The written test isn't all that complicated—especially if you've already passed your No-Code—and you only have to send and receive Morse code at five words per minute. Figure 7-10 is a Morse code chart.

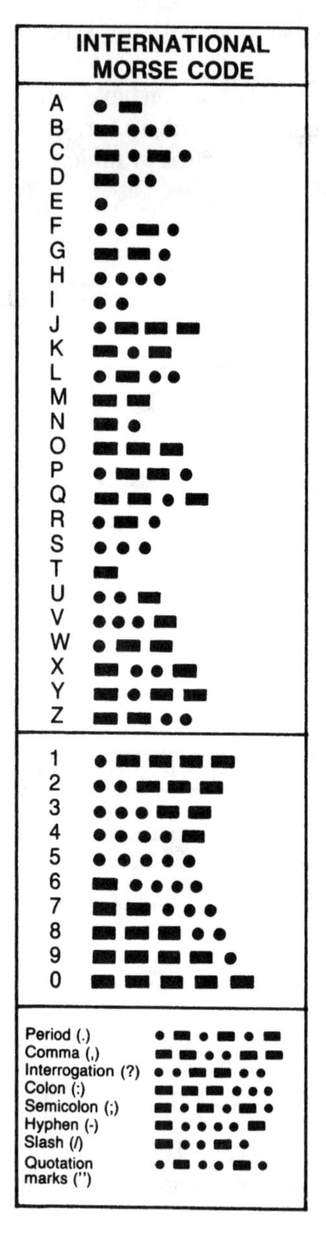

7-10 An International Morse Code chart. ARRL.

While most Novice shortwave frequency privileges are for code-only transmission, you are allowed to use voice on the upper portion of the 10 meter band (28.300–28.500 MHz). When skip conditions are right, you can communicate with operators hundreds—even thousands—of miles away.

Writing ham radio reception reports

When you fill out a QSL card, you will be reporting on the quality and strength of the other station's signal. Ham radio operators use the RST—Readability/Signal/Tone—system to evaluate Morse code (CW) transmissions (Table 7-1). For voice operation, you only need to evaluate readability and signal strength.

Table 7-1. The RST (Readability/Signal/Tone) signal reporting system.

	Readability
1	Unreadable
2	Barely readable
3	Readable with considerable difficulty
4	Readable with practically no difficulty
5	Perfectly readable
	Signal strength
1	Faint signal—barely perceptible
2	Very weak signal
3	Weak signal
4	Fair signal
5	Fairly good signal
6	Good signal
7	Moderately strong signal
8	Strong signal
9	Extremely strong signal
	Tone (for code transmissions only)
1	Extremely rough, harsh and broad tone
2	Very rough, harsh tone
3	Rough, rippling tone
4	Moderately rough, rippling tone
5	Moderate, with some rippling sound
6	Moderate, hardly any rippling in tone
7	Near pure tone, only traces of rippling
8	Near perfect tone
9	Perfect tone

The next class of license beyond Novice is General. As you can see from the ham radio frequency chart, you get some voice privileges on all ham bands when you can pass the more difficult written test and 13-word-per-minute code exam.

Study materials are available by mail for all classes that prepare you for FCC examinations. The National Amateur Radio Association has books and computer software to take you all the way up to an Extra class license (Fig. 7-11). For information, call the Association at 1–800–GOT–2–HAM.

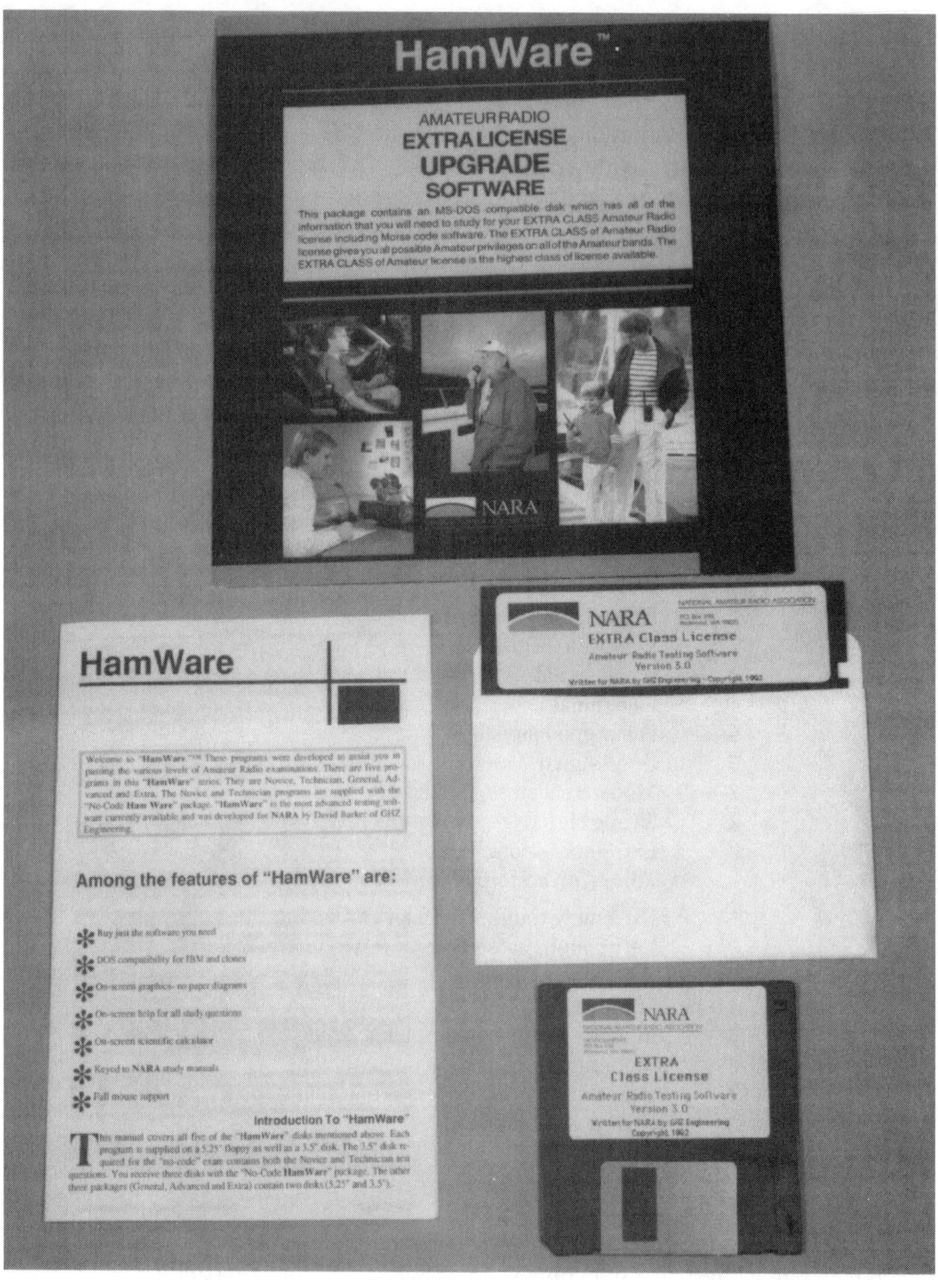

7-11 The National Amateur Radio Association has study materials available for all classes of amateur radio licenses. NARA photo.

Q signals

Hams—especially when they're transmitting in code—use Q signals to save time. Over the decades, dozens of Q signals have developed. Some are even used in shortwave. Table 7-2 lists the Q signals most commonly used in ham radio.

Table 7-2. International Q signals.

QRL	Is this frequency busy?
QRM	Interference from other stations.
QRN	Atmospheric noise (static).
QRO	Increase transmitter power.
QRP	Lower transmitter power.
QRS	Send (code) slower.
QRT	Stop transmitting.
QRZ	What are your call letters?
QSA	What is my signal strength?
QSB	Your signal is fading.
QSL	Did you copy? (Also means please send a QSL card.)
QSO	Contact (Two-way transmission).
QSY	Change frequency to ______.
QTH	Location (usually city and state).

ITU phonics

The International Communications Union (ICU) has developed a list of phonics, which are used worldwide to help voice operators get their messages across under difficult operating conditions. When band conditions are bad and you can barely hear the operator you're in contact with, phonics can make the difference between hearing and not hearing what's being said. Table 7-3 is a list of ITU Phonetics, or Phonics.

Table 7-3. ITU phonetics.

A	Alfa	I	India	R	Romeo
B	Bravo	J	Juliett	S	Sierra
C	Charlie	K	Kilo	T	Tango
D	Delta	L	Lima	U	Uniform
E	Echo	M	Mike	V	Victor
F	Foxtrot	N	November	W	Whiskey
G	Golf	O	Oscar	X	X-ray
H	Hotel	P	Papa	Y	Yankee
		Q	Quebec	Z	Zulu

Morse code shortcuts

Hams that communicate in Morse code often use abbreviations. It saves them the trouble of spelling out every word they use—and gets the message across a lot faster! Table 7-4 is a list of the most commonly used abbreviations.

Table 7-4. Abbreviations for Morse code operators.

ABT	About
ADR	Address
AGN	Again
ASCII	American National Standard Code for Information Interchange
BCI	Broadcast (shortwave) interference
BCNU	Be seeing you (good by)
BK	Back—or—break-in
C	Yes
CFM	Confirm
CL	Closing, signing off the air
CQ	Calling any station
CUL	See you later
CW	Continuous wave—Morse code
DR	Dear (used as a greeting by many overseas hams)
DX	Distance
ES	And
FB	Fine business (great)
FREQ	Frequency
GA	Good afternoon—or—go ahead
GE	Good evening
GM	Good morning
GN	Good night
HI	Laughter in CW
HR	Here
HV	Have
HW	How
ITU	International Telecommunications Union
LID	A bad operator
N	No
NCS	Net control station
NR	Number
NW	Now
OM	Old man—refers to another operator until names are known
PSE	Please
R	Received
RCVR	Receiver
RFI	Radio frequency interference
RPT	Repeat
RTTY	Radio teletype
SASE	Self-addressed stamped envelope
SIG	Signal
SKED	Schedule
SRI	Sorry
SSB	Single side band
SWL	Shortwave listener
TNX	Thanks
TVI	Television interference
VFO	Variable frequency oscillator

Table 7-4. Continued.

VY	Very
WA	Word after
WB	Word before
WPM	Words per minute
WX	Weather
XMTR	Transmitter
XTAL	Crystal
XYL/YF	Wife
YL	Single lady
73	Best wishes
88	Love and kisses

Transceivers for the shortwave ham bands

Transceivers for the high-frequency (shortwave) ham bands look quite a bit different than 2 meter hand-held units. Since they are more powerful (most run about 100 watts, as opposed to a hand-held's power of around 5 watts) and have to be able to operate in several different frequency ranges, it takes a larger and heavier piece of equipment to hold it all.

Figure 7-12 is an ICOM IC-R726 transceiver. It operates on all high-frequency bands and the 50 MHz VHF band. Like nearly all modern transceivers, it has digital frequency readout and operates in both code and voice modes.

7-12 Icom IC-R726 transceiver. Icom photo.

The Realistic HTX-100 10 Meter Transceiver (Fig. 7-13) makes it possible for you to enjoy your hobby while you're on the road. It transmits in both voice and Morse code, and has a selectable power output of either 5 or 25 watts.

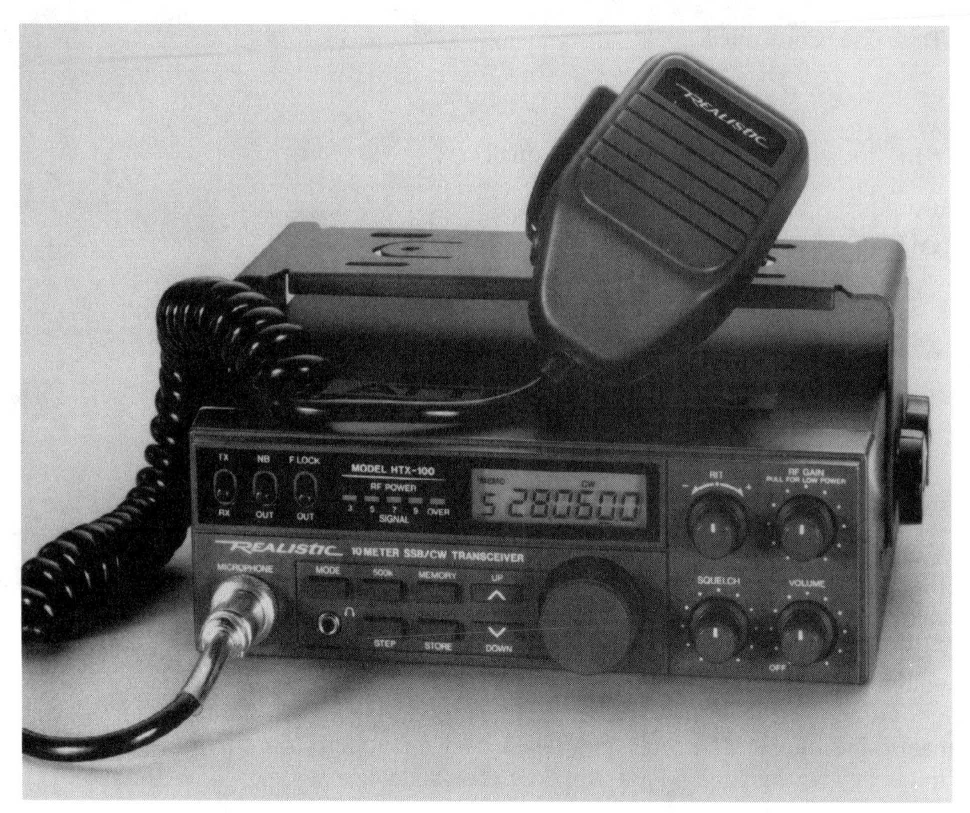

7-13 Realistic HTX-100. Radio Shack.

Antennas

The wavelengths you'll be using on the shortwave ham bands are much longer than what you used with your 2 meter transceiver. And when you're dealing with waves 40 and 80 meters long, as you will be on these bands, you'll have to put up an external antenna to get your signal out. The three most commonly used amateur antennas are:

- *Dipole antennas.* They're made from two long pieces of wire (cut to the right frequency), two insulators, and a lead-in cable. Trap dipole antennas have two cylinder-shaped traps that "fool" the radio waves into behaving as they would if the antenna were cut to their length. So you can use them on several different frequency bands.
- *Vertical antennas.* These antennas, made from a piece of metal mounted vertically, are great when you have limited space. They sometimes come with traps, so you can use them on all high-frequency (shortwave) bands, and they can be ground- or pole-mounted.
- *Beam antennas.* These are made from several thin rods of metal that transmit and guide your signal in one direction. Usually, they're mounted on a tower

with a rotor, so you can aim your transmission at any part of the world you wish. Most beam antennas are designed for the 2, 6, 14, 21, and 28 MHz bands. Figure 7-14 shows how they look.

Nearly all ham radio stores and mail-order catalogs sell prefabricated antennas you can assemble. But when you're ready to put a high-frequency ham station on the air, it's a good idea to work with a local ham who has some experience in putting up antennas, making sure they're safe against lightning strikes and so on.

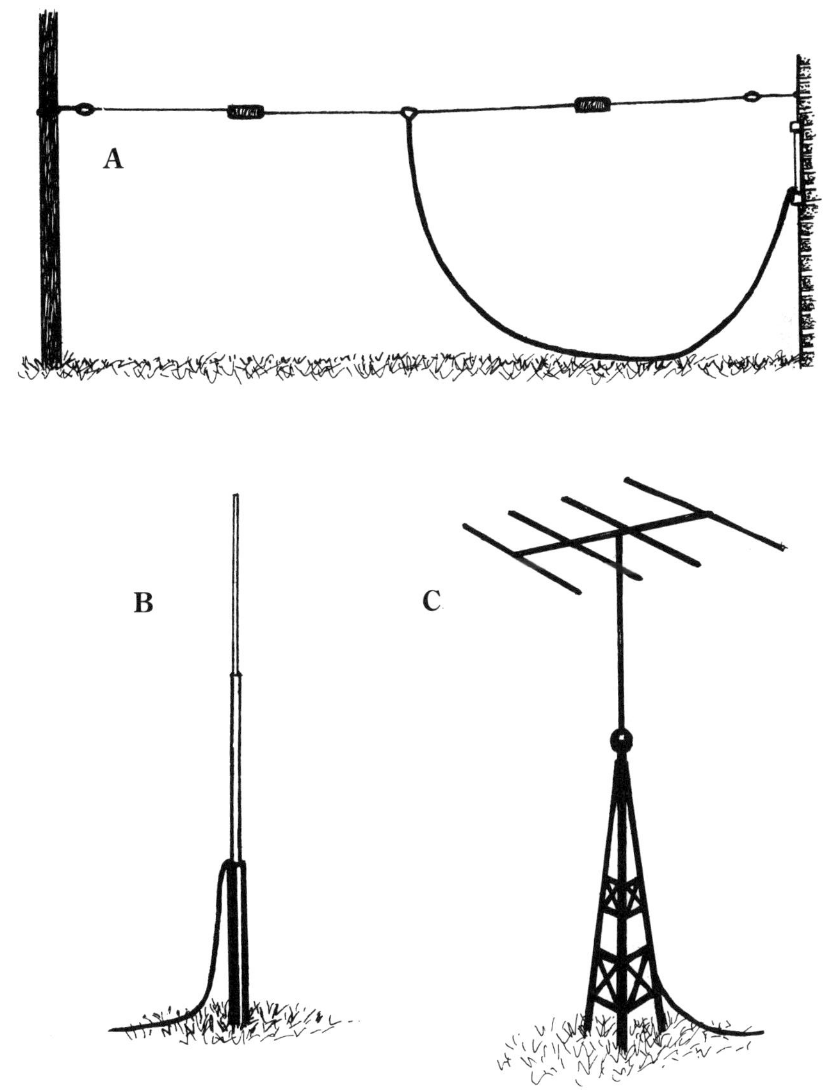

7-14 (A) trap dipole, (B) vertical, and (C) beam antennas are popular with ham radio operators.

Space shuttle transmissions on the ham bands

Several amateur radio clubs across the country broadcast retransmissions of communications from NASA space shuttle missions. You can listen in as astronauts, over 200 miles above us, talk to ground control stations as they orbit the earth.

The clubs participating in this program are:

WA3NAN
Goddard Space Flight Center
Greenbelt, Maryland:
3.860, 7.185, 14.295, and 21.395 MHz.

W5RRR
Johnson Space Center
Clear Lake City, Texas:
3.850, 7.227, 14.280, 21.350, and 28.495 MHz.

W6VIO
Jet Propulsion Lab
Pasadena, California:
3.840 and 21.280 MHz.

All transmissions are in single sideband, so you'll need an SSB or BFO control on your receiver in order to hear them.

Figure 7-15 is a QSL card from WA3NAN, the Goddard Space Flight Center Amateur Radio Club in Greenbelt, Maryland, which is used to retransmit shuttle communications. The QSL shows the Solar Max Satellite, which was repaired during a space shuttle mission.

If you're licensed and own a 2 meter transceiver, you'll have the opportunity to talk to ham/astronauts as they cross over your part of the world. The downlink (space-to-earth) frequency is 144.55 MHz, and the uplink (earth-to-space) frequencies are 144.91, 144.95, and 144.97 MHz.

Schools use ham radio to spark students' interests

More and more frequently, teachers are discovering that ham radio, as well as shortwave listening, is a great way to spark young people's interest in school work. Joe Fairclough, an English teacher at New York City's Junior High School 22 (Fig. 7-16), has been using ham radio as a teaching tool since 1979. With the help of donated radio equipment, his "Education through Communication" program, now known as EDUCOM, has proven to be a success in keeping kids in school and off the streets. EDUCOM has now become a nationwide, nonprofit group, dedicated to promoting ham radio as a teaching tool.

The Goddard Space Flight Center (GSFC) was named in honor of Dr. Robert H. Goddard, the pioneer of American Rocketry, who was also a radio amateur. GSFC is located in Prince Georges County, Maryland, Northeast of Washington, D.C. The center performs basic space research, is responsible for the development of unmanned satellites and manages/operates NASA's Global Tracking Network, including the Tracking Data Relay Satellite System (TDRSS). The Goddard Amateur Radio Club is an employee recreational organization. Please feel free to contact us whenever you visit the center.

The Solar Max Satellite, developed by GSFC and repaired during Shuttle flight 41C is depicted on the face of this card. This satellite typifies the center's commitment to scientific research, engineering expertise and spacecraft serviceability in the shuttle era.

Radio KA8KGI Confirms Your ~~QSO~~/Reception as Follows:

DATE	TIME (GMT)	REPORT	FREQ.	MODE
12-2-83	0342	Thanks Anita	3.860	SSB

QSL via WA3NAN, Box 86, Greenbelt, Maryland, U.S.A. 20770

73, Bob NP4B OPR

Color QSL By K2RPZ Print, Box 412, Rocky Point, NY 11778
96286-D

7-15 WA3NAN retransmits space shuttle communications.

Here's how it works:

- Morse code is taught at the beginning of the school year. Later, vocabulary and spelling drills are done in Morse.
- Ham radio study material is used as an English textbook. Grammar, sen-

tence structure, and so on are learned as the students prepare to pass their FCC exams.

- Reading assignments come from amateur radio magazines, newsletters, and other radio-related publications.
- Students learn to compose letters by writing to ham radio operators they contact on the air.

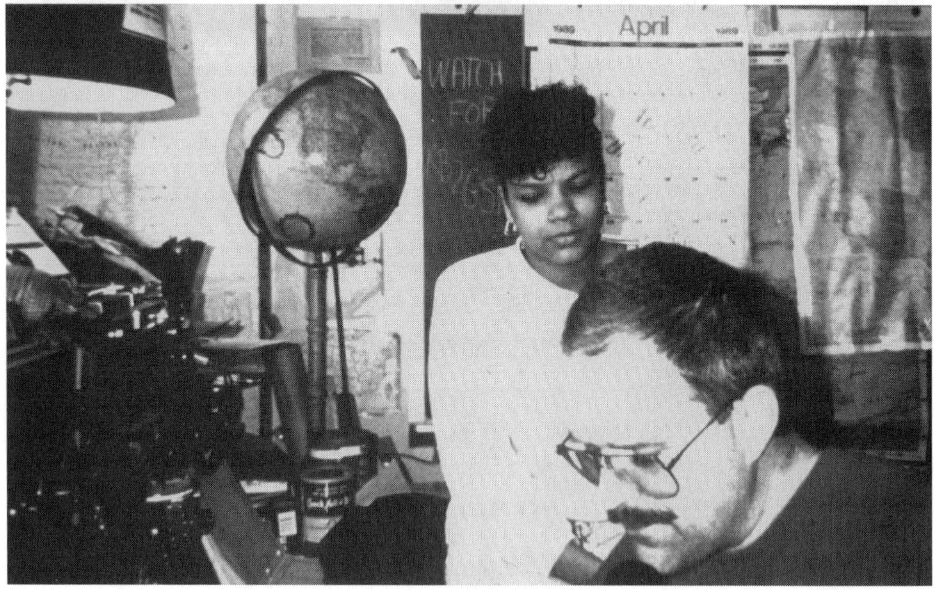

7-16 Joe Fairclough, along with Nicol—KA2GGW, and Ralph—KB2HUR at the WB2JKJ club station.

Currently, the schedule for EDUCOM's educational network is:

7.238 MHz, Monday through Friday, 7:00 to 8:30 A.M. Eastern
21.395 MHz, Monday through Friday, 9:00 A.M. to 3:00 P.M. Eastern

Figure 7-17 is a QSL card from Jr. High 22's club station—WB2JKJ.

7-17 A QSL card from WB2JKJ.

Ham radio magazines and newsletters

If you've decided to become a ham, magazines are an excellent resource. They tell you about the activities going on in all fields of amateur radio, keep you informed about changes in FCC rules and regulations, and carry advertisements of transceivers, antennas, and other products you'll want to purchase for your station. All ham radio magazines take their own unique approach to the hobby. *The Amateur Radio Communicator* (Fig. 7-18), for example, is written primarily for new operators or people who are preparing for the FCC exam. Others, including *CQ*, *QST* (Fig. 7-19), and *73*, contain material for hams of all levels. They feature projects to build, information on DX contests, new antenna designs, proposed changes in FCC rules and regulations, and information on the ham radio hobby in general.

Some publications, such as *The Spec-Com Journal*, *Amateur Television Quarterly*, *USATV*, and *AMSAT*, focus on special operating techniques, including slow-scan TV, fast-scan TV, satellites, RTTY (Radioteletype), and the use of computers in packet amateur radio communication. *Radio News* (Fig. 7-20), published by Icom America,

7-18 *The Amateur Radio Communicator* magazine, published by the National Amateur Radio Association, 1-800-GOT-2-HAM.

helps new hams learn about all aspects of the hobby and showcases Icom products. Here's how to get in touch with some ham publications:

- *The Amateur Radio Communicator*
  1645 Redmond Way
  Suite 232
  Redmond, WA 98052
  (800) GOT–2–HAM

Focuses mostly on introducing people to ham radio. Gives basic, easy-to-understand information on all aspects of the hobby.

- *Amateur Television Quarterly*
  540 Oakton Street
  Des Plaines, IL 60018–1950
  (Amateur TV magazine.)

- *AMSAT*
  PO Box 27
  Washington, DC 20004
  (301) 589-6062

Explains how you can transmit signals around the world on ham radio satellites.

- *CQ Magazine*
  76 North Broadway
  Hicksville, NY 11801

7-19 *QST*, the monthly magazine of the American Radio Relay League.

YOUR PERSONAL GUIDE TO ENJOYING AMATEUR RADIO.

VOLUME 4 ISSUE 2 1991-92

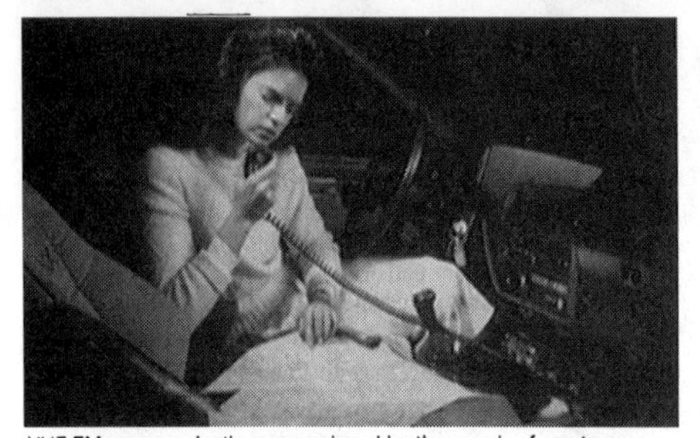

VHF FM communications are enjoyed by thousands of amateurs throughout the U.S. and the world . They are a great traveling companion and priceless emergency aid.

Amateur radio cosmonauts aboard the Soviet space station MIR often contact U.S.-based amateur on 145.55 MHz. They use an Icom IC-228A two-meter transceiver.

VHF/UHF Activities: Unlimited Fun!

Hold onto your microphone, let your eative thoughts flourish, and get ready r some amateur radio excitement. This ecial edition of Icom's RADIO NEWS loaded with terrific information and lpful guidance for all amateurs from de-free Technician to Extra Class ensees. It highlights the fascinating ea of VHF/UHF communications and eir numerous activities you can enjoy ght now, and they are indeed extensive! Heading the list in popularity is two eters/144 MHz with its diverse FM, peater, OSCAR satellite, packet and B/CW activities. Close on its heels are e 70cm/440 MHz and 23cm/1.2 GHz nds and these upper frequencies are o an FM and satellite enthusiast's haven. The 1.25-meter/222 MHz and six-meter/ MHz bands are also becoming favorite eeting grounds among an increasing mber of amateurs each day. The previous erview was only a starter!

FM and repeater activities on today's VHF/UHF bands are complimented with an unbelievable array of compact, deluxe featured transceivers you can set up anywhere and use with small antennas. These rigs are perfect for amateurs living in antenna-restricted neighborhoods or apartments, plus they are a great traveling companion. Our VHF and UHF bands are truly the wave of amateur radio's future, and this newsletter's inside pages give you full details for joining the action in high style! **Icom is on your side and dedicated to your amateur radio enjoyment!**

Some amateurs are reading Icom's RADIO NEWS for the first time, especially new codeless Technician licensees. Icom congratulates you on passing the exam, and welcomes everyone aboard. If you wish to receive future newsletters, drop us a letter.

Incidentally, Icom encourages you to practice your five wpm Morse and add worldwide 10-meter SSB/CW enjoyment to your radio fun. If you are a Novice licensee, study and pass element three so you can also work two-meters. The final step into HF radio action then simply involves increasing your code copy speed to 13 wpm for a General class license.

ICOM FIRST IN SPACE

Beginning January 1991, the Soviet-manned space station MIR (call sign U2MIR) became active on two-meters with an Icom IC-228A FM transceiver and packet TNC plus laptop computer for packeting. Soviet space coordinator, UW3AX, said Icom was selected because of its compact size and reliable operation. Musa, UV3AM, and future Soviet cosmonauts will use this equipment for working amateurs worldwide during off hours. Listen to them on 144.55 MHz.

Published by: Icom America, Inc.

7-20 *Radio News*, published by Icom, is helpful to new hams.

DXing, contests, awards, construction projects, antennas, and operating in general.

- *Digital Digest*
  Avro & Associates
  4063 Goldenrod Road
  Winter Park, FL 32792

For people who want to use their computer with their amateur radio station.

- *Nuts and Volts*
  T & L Publications Inc.
  PO Box 111
  Placentia, CA 92670

Electric parts and used equipment for hams, experimenters, and computer buffs.

- *QST*
  225 Main Street
  Newington, CT 06111

Information for hams in all levels of the hobby—contests, DXing, and construction projects. Special sections on activities of the American Radio Relay League.

- *Radio News*
  Icom America, Inc.
  2380 116 Avenue
  Bellevue, WA 98004

Intended mostly for people new to the hobby who would like to know more about what they can do with amateur radio equipment.

- *73 Amateur Radio*
  70 Route 202 North
  Peterborough, NH 03458

Information, construction projects, contests, and DX tips for hams of all levels.

- *The Spec-Com Journal*
  PO Box 1002
  Dubuque, IA 52004

Information on ham radio TV, satellite communication, RTTY, and Packet Radio.

- *Tucson Amateur Packet Radio*
  PO Box 12925
  Tucson, AZ 85732
  (602) 749–9479

Specializes in "packet" computer communications via ham radio.

- *USATV*
  1520 Cerro Drive
  Dubuque, IA 52001

Amateur TV magazine.

- *Westlink Report*
  28221 Stanley Court
  Canyon Country, CA 91351
  (800) HAM–7303

Newsletter on all aspects of ham radio, published 26 times a year. Covers FCC decisions, new equipment reviews, propagation forecasts, and news of ham radio in general.

- *W5YI Report*
  PO Box 565101
  Dallas, TX 75356-5101
  (817) 461–6443

Up-to-the-minute news about the ham radio hobby, published every two weeks.

Monitoring Times and *Popular Communications* feature monthly columns on ham radio (see Grove Enterprises, Inc., and *Popular Communications* in Appendix C). Figure 7-21 shows *Monitoring Times* editor, Bob Grove, at his ham radio station.

7-21 *Monitoring Times* editor Bob Grove at his ham radio station.

8
The mysterious world of radio waves

INFORMATION OF ALL TYPES IS TRANSMITTED BY RADIATION. SOUND WAVE radiation carries your voice across the room. Light wave radiation brings us information about objects in our environment, the sun, planets, and even distant stars.

Radio waves also transfer information. They bring news, talk, entertainment, and music through the air at the speed of light, from the transmission (radiation) tower to your receiver. Figure 8-1 is a chart of the radio frequency spectrum.

Exactly what is a radio wave?

All radio waves are made up of electric and magnetic fields. The wavelength, frequency, and strength of the magnetic field is always the same as that of the electric field (Fig. 8-2). A radio wave begins its life at the transmitter, where electric charges are switched from positive to negative so quickly that waves of electromagnetic energy are created, then pushed off the antenna and into the air.

Radio waves get the message across

To transfer information by means of a radio wave, you have to have some way of altering it so it will be able to carry your message from one place to another. Here are three methods:

Morse code

Morse code uses the simplest method of altering a radio wave. When you press the code key down, you transmit waves. When you release the key, the transmission ends.

In Morse code, each letter, number, and punctuation mark is made up of a different combination of long and short taps on the key. For many decades, ham

Radio Frequency Spectrum

Between the very low frequencies, which can be heard as sound, and the very high frequencies which can be seen as light, are the very useful frequencies we use for radio communications. Although the exact limits are only roughly defined and subject to change as our technology advances, the most usable frequencies are clustered in the center of this radio spectrum. The bar graph below illustrates radio's position in the frequency spectrum.

Sound to Light Spectrum

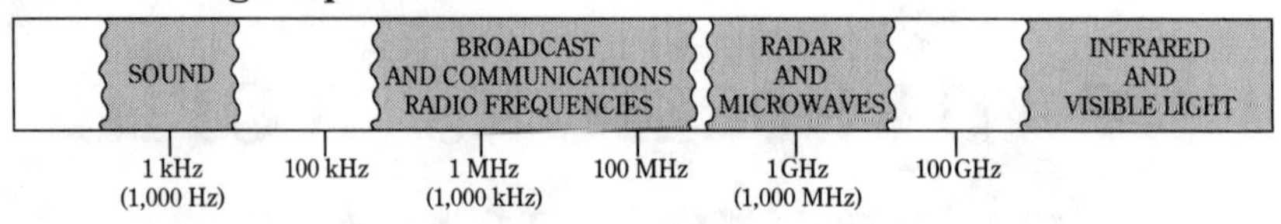

International and Federal Communication Commission Regulations have designated certain portions of the radio spectrum for various services, from TV broadcasts to two-way police communications. The graphs below give an expanded view of the spectrum from 0.5 to 500 megahertz and show some of the principal FCC assignments.

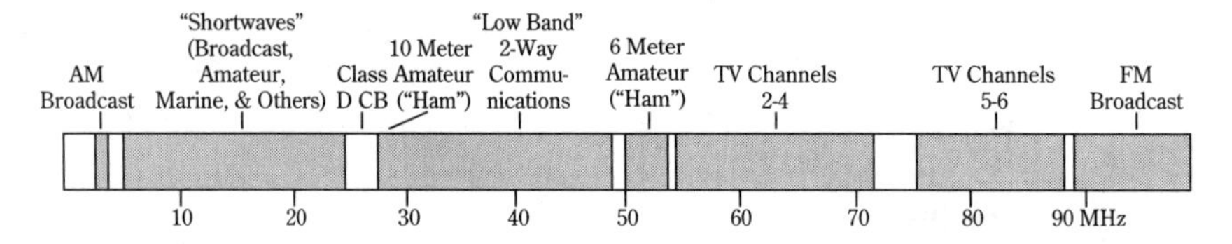

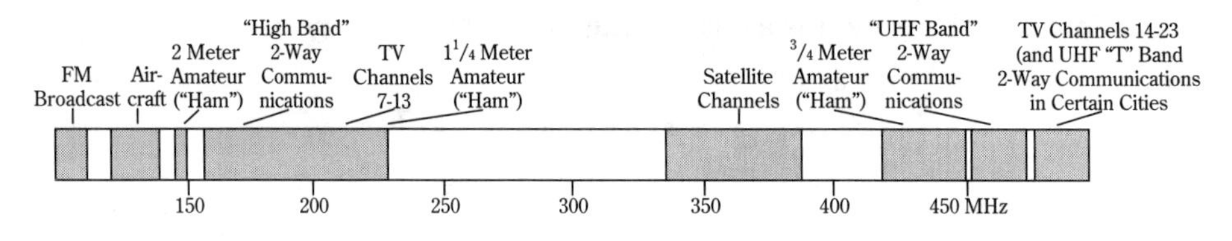

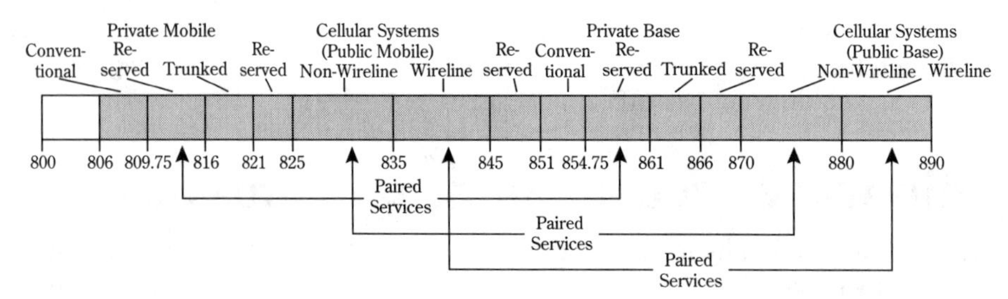

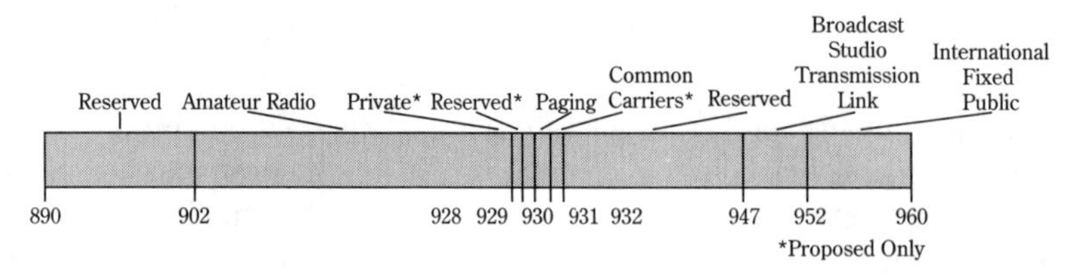

8-1 This chart shows the radio frequency spectrum. Courtesy of the Bearcat Radio Club, Kettering, Ohio.

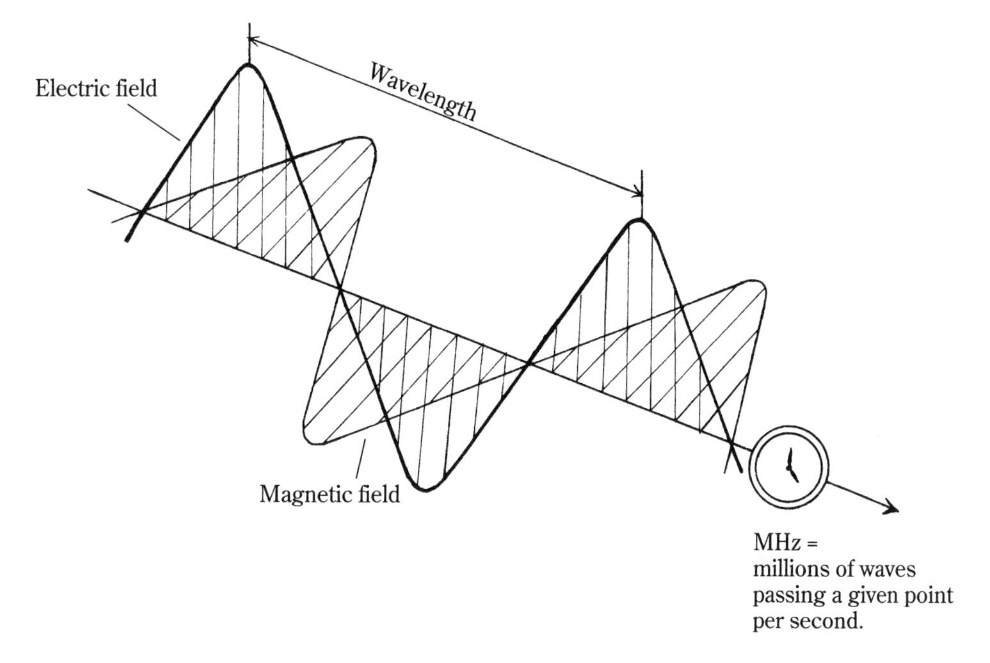

8-2 Radio waves are made up of an electric and a magnetic field.

radio operators around the world have been using Morse code to communicate. Anyone that knows the Morse system can tell you exactly what another operator is saying with his "dits" and "dahs."

But even though it is the simplest way to transmit messages, Morse code does have its drawbacks. It takes a tremendous amount of time to spell out each word. You can't transmit voices or music with Morse code, and very few people in our society know how to read it.

Amplitude Modulation (AM)

The oldest method of transmitting voice and music through the airwaves is by amplitude modulation. (Modulate means to change a wave so it can contain and carry information.) This is accomplished by combining a sound wave from a microphone, tape, record, or CD with a "carrier" radio wave. The result: a wave that transmits voice or programming as its amplitude (intensity) increases and decreases. Amplitude modulation is used by stations broadcasting in the AM band and by most international shortwave stations.

Single sideband, a modified form of amplitude modulation, takes up less band space than a regular AM signal. A sideband is eliminated—then replaced by the receiver (if it has a BFO/SSB control) when you tune it in. Single sideband is used extensively by utility stations, ham stations, some radio pirates, and a few international shortwave broadcast stations.

Frequency Modulation (FM)

Another way to convey information, voice, and music on a radio wave is to slightly change, or modulate, the frequency. The main advantage FM broadcasting is of it is static-free. But the drawback to FM is since the frequency is varied, stations take up more room on the band.

Frequency modulation is, of course, used on the FM band. And it is used for "action band" and ham transmissions in the VHF/UHF frequency range.

How radio waves travel

Once radio signals leave the antenna, they rush through the air at the speed of light. Some waves stay close to the ground, while other waves travel up into space. The frequency (number of waves passing a given point per second) and wavelength (the length of a radio wave from one energy peak to another if you could hold it still and measure it) determine where you will hear the transmission on your dial and how far away the signal can be heard.

Ground waves

A ground wave bends along the curve of the earth until it becomes too weak for your radio to receive it (Fig. 8-3A). On the AM and lower shortwave bands, the ground wave can cover many miles.

Skip waves

On all the shortwave bands, and on the AM band at night, radio wave "skip" can carry transmissions incredible distances. Once the signal reaches the ionosphere, it is bent and returned to earth, hundreds, if not thousands, of miles away (Fig. 8-3B).

During the evening, lower shortwave frequencies have the best skip conditions while daytime "skip" propagation benefits stations on the higher end of the shortwave frequency spectrum. Ham radio operators know that even low-power QRP stations running 10 watts of power or less can often make transoceanic contacts on frequencies above 14 MHz.

Line-of-sight transmissions

On VHF, UHF, and microwave frequencies, signals usually travel in line of sight. In other words, radio waves at frequencies over 50 MHz usually travel in a straight path and will not curve over the horizon (Fig. 8-3C). That is why it is so important for fire departments, police stations, and other users of VHF/UHF frequencies to get their transmission tower as high as possible.

As a general rule, remember that the higher the frequency of a radio wave, the more the wave behaves like waves of light. Light waves travel in straight lines. (If you remember the pictures our astronauts took on the moon, shadows were completely black. Shadows aren't nearly as dark on earth because they hit the molecules in our atmosphere and move in different directions. The moon has no atmosphere to deflect light waves—thus, shadows are extremely black.)

Freak conditions cause VHF skip

Unusual weather or ionospheric conditions can bring in VHF signals from several states away.

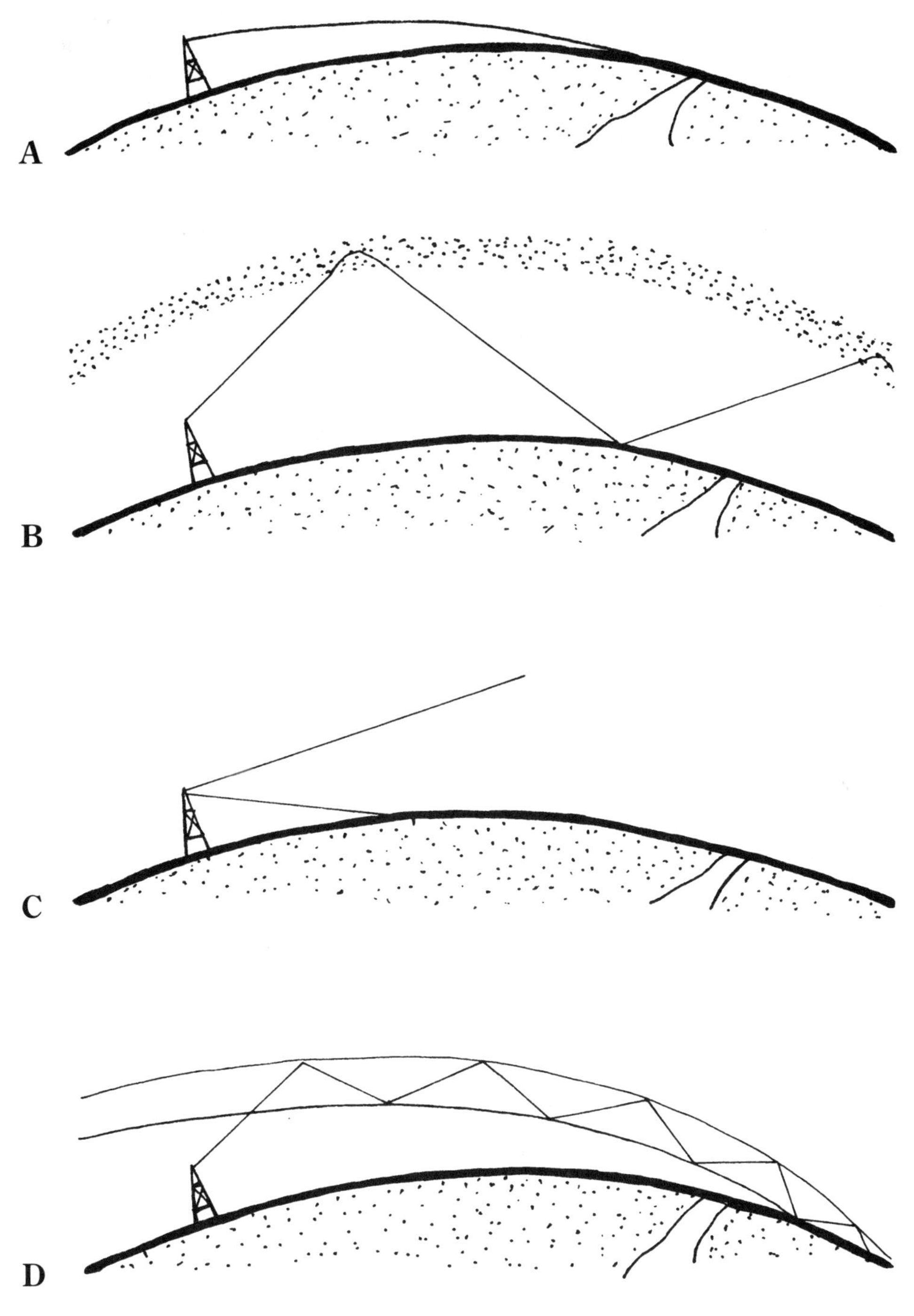

8-3 (A) Ground wave, (B) skip wave, (C) line-of-sight wave, (D) tropo skip wave.

Tropospheric ducting

One of the most common forms of skip on "action band" scanner frequencies, FM radio, VHF ham bands, and television is tropospheric ducting. It takes place during temperature inversions and is most likely to occur in the spring and summer months, miles away. When a tropospheric duct is formed, normally line-of-sight waves get caught in the duct and are transported to listeners hundreds of miles away (Fig. 8-3D).

On many occasions, signals that arrive by means of tropospheric ducting can be strong enough to interfere with local broadcasts. While you probably won't be able to get good enough reception to identify a skip station that's broadcasting on top of a local station, your best bet is to tune to nearby frequencies (or nearby channels, in the case of TV) and try your luck there.

When a tropo skip occurs, several stations often try to pop in at once on the same TV channel. Sometimes you'll hear the sound from one station while viewing thc picture from another.

The better and higher your antenna is, the more skip stations you'll be able to pick up. But if your space is limited, don't worry. Even rabbit ear antennas can bring in skip transmissions. (If you have cable service, you'll have to disconnect it from your TV—or you won't be able to see *any* skip.)

Figure 8-4 is a letter from TV station CKND in Manitoba, Canada. It was picked up in West Virginia by means of a normal rooftop TV antenna.

Sporadic E skip

Strongly ionized patches can develop in the E-layer of the ionosphere 20 to 40 hours after a major solar flare. They are most common in May/June and again around December. Sporadic E patches are dense enough to affect radio waves transmitted on the upper portions of the shortwave band, the FM, and TV bands, as well as lower VHF scanner frequencies, bending and returning them to earth (Fig. 8-5).

Sporadic E skip can disrupt police and fire and ambulance communications and bring in TV stations from hundreds of miles away. It also helps ham radio operators on the 6 meter band reach out across the country on low-power, hand-held transceivers.

Radio signals in space

As soon as a radio signal leaves the antenna, it is, for all practical purposes, traveling in space. Depending on the frequency and ionospheric conditions, sky waves can either be absorbed by the D-layer, "skip" off the E- or F-layer and return to earth, or travel out through the ionosphere into space.

The highest frequency you can use to send radio "skip" waves from one part of the planet to another is known as the MUF, or maximum usable frequency. If you use a frequency higher than the MUF, the signal will pass through the ionosphere and never be heard by earthlings again!

The ionosphere surrounds our planet

As you can see in Fig. 8-6, the layers of our ionosphere are different on the night side of our planet than on the day side. The part of the earth that is facing the sun has a

CKND

Channel 9 Winnipeg/Channel 2 Westman

August 21, 1987

Anita L. McCormick

Dear Ms. McCormick:

Thank you for your letter dated June 6/87. I am happy to report to you that you did receive our station on TV Channel 2 on June 4 at 6:30 p.m..

Sincerely,

Robin Schreiber

Robin Schreiber
Promotion Department
CKND-TV.

603 St. Mary's Road, P.O. Box 60, Winnipeg, Canada R2M 4A5
Telephone: (204) 233-3304 Telex: 07-55270 Telecopier: (204) 233-5615
Owned & Operated by CanWest Broadcasting Ltd.

8-4 A letter from TV station CKND.

D-, E-, F1-, and F2-layer. But the side of the earth turned away from the sun and its endless stream of radiation has only two layers of ionization—the E-layer and an F-layer.

On the dark side of our planet, the E-layer doesn't receive sufficient radiation to remain ionized. And when evening comes, the F1- and F2-layers drift together to form one F-layer.

That's why day- and nighttime skip conditions differ so much.

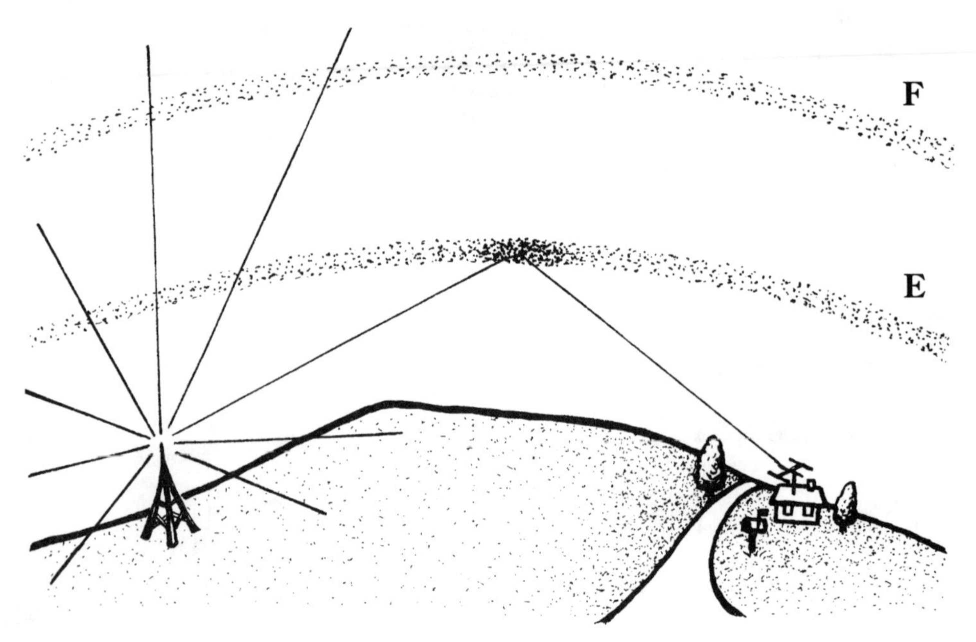

8-5 Sporadic E skip gives long-distance reception on VHF and upper shortwave band frequencies.

The sun's role

The radio signals we enjoy originate on earth. But it is the sun that supplies us with life-giving warmth and also creates the means for radio waves to "skip" around the planet. Massive amounts of radiation from the sun cause electrically charged layers of gas to form in the upper layer of our atmosphere—and make radio wave propagation possible.

The sunspot cycle

Every 22 years, the sun's magnetic poles completely reverse. And as the cycle approaches the halfway point (approximately every 11 years), the number of spots on its surface dramatically increases. At the high point of the cycle, over 100 spots are visible at the same time.

Sunspots are areas of extremely strong magnetic activity. In fact, the magnetic energy contained in a sunspot is so great that it forces heat away, keeping spots somewhat cooler than the surrounding area. (That's why they appear dark by comparison.)

Solar flares

Solar flares are almost always present in areas of high sunspot activity. They erupt as a result of intense, twisted magnetic fields between groups of sunspots (Fig. 8-7). When a large solar flare hurls radiation out into space towards our planet, it can disrupt long-range AM and shortwave communications for hours, or even days.

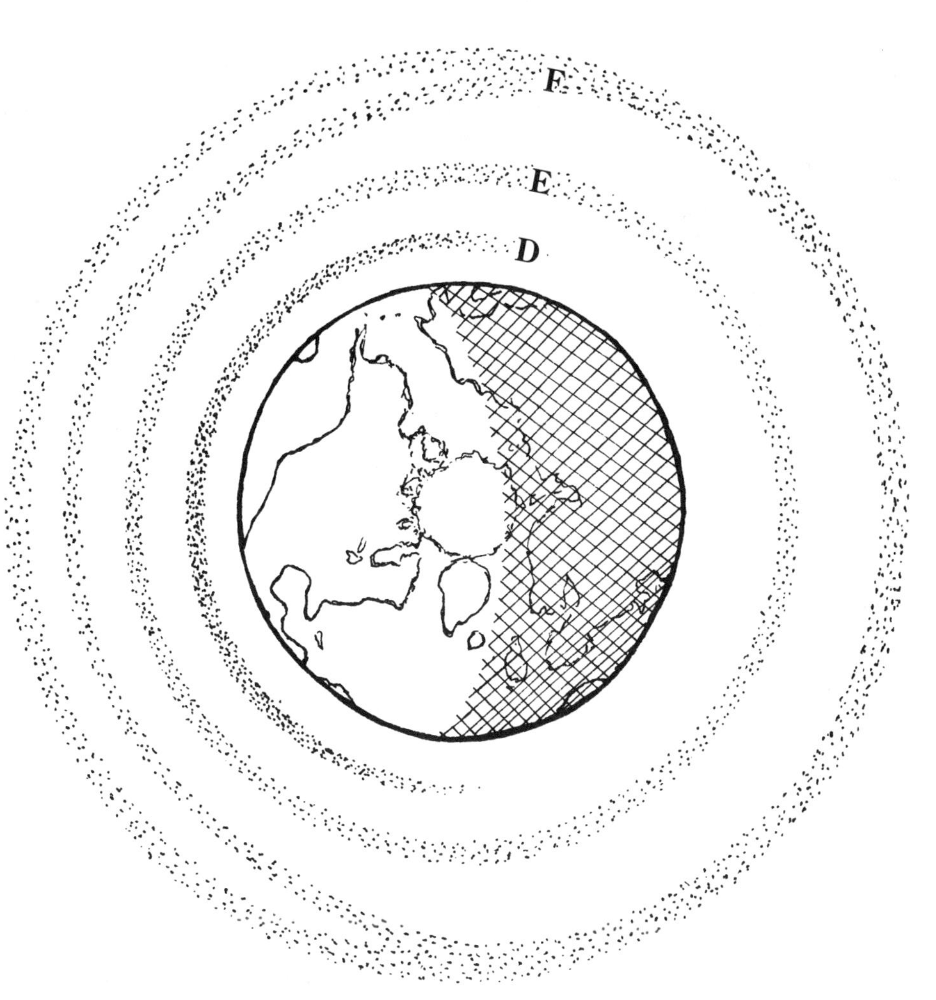

8-6 As the earth rotates, it passes through different ionospheric conditions.

The magnetosphere

As solar radiation rushes towards earth, it encounters our magnetosphere, two doughnut-shaped magnetic fields that surround the planet and shield us from harmful radiation (Fig. 8-8). When radiation, from the sun or from outer space, enters the magnetosphere, it bounces up and down through the powerful magnetic fields, gradually losing its energy. Finally, the particles of radiation fall down into the atmosphere, usually near the poles, and create beautiful auroras.

Geomagnetic storms

As long as the normal amount of radiation enters the magnetosphere, this arrangement works out fine. But when major solar flares occur, we have what is known as a geomagnetic storm. The magnetosphere can only handle so much radiation at one

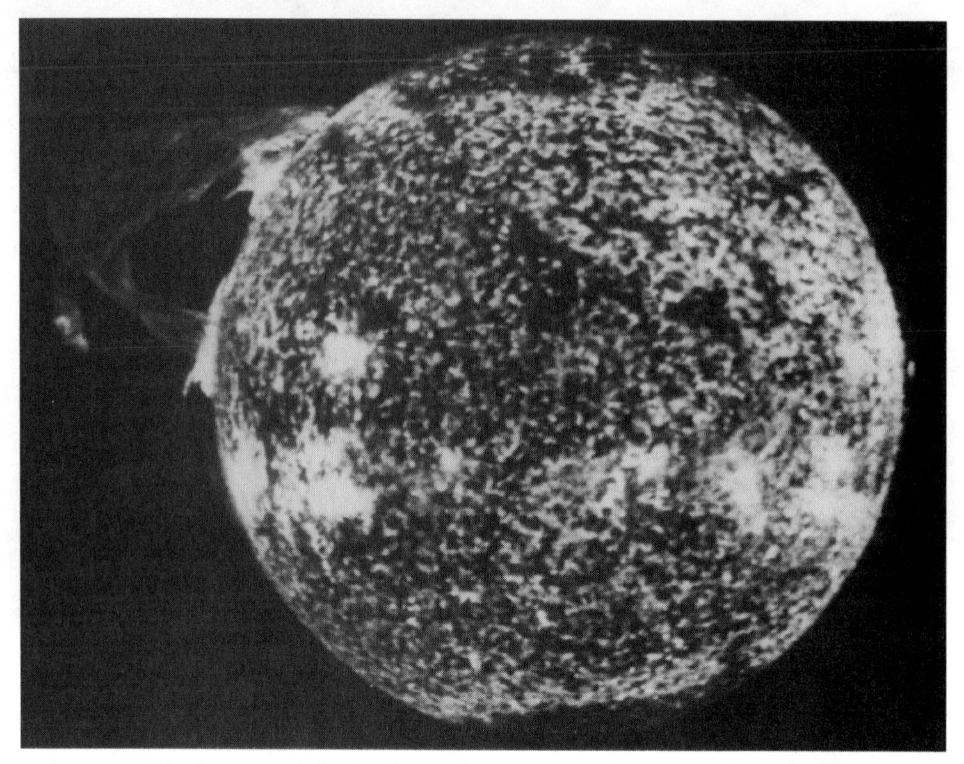

8-7 Sunspots and solar flares affect reception here on earth. NASA photo.

time. Solar particles fall out of the magnetosphere before their energy has time to be absorbed. This has an adverse effect on our ionosphere and makes long-distance shortwave communications extremely difficult, for hours, sometimes even days.

During a geomagnetic storm, the D-layer of our ionosphere is charged to the point that it absorbs and destroys radio signals at a much higher rate than usual. Even the upper layers of the ionosphere behave differently. The visible evidence of a geomagnetic storm are brightly colored auroras in the night sky, much closer to the equator than we can usually see them.

Solar flux index

The solar flux index is a way of measuring solar activity—and the effect it has on radio communication. The solar flux for the day is transmitted on time station WWV at 18 minutes past each hour. As the number of sunspots rises, so does the solar flux.

At the bottom of the solar cycle, when few sunspots are visible, the solar flux is generally between 65 and 85. Midway through the cycle, the number is 85–150. And at the high point of the cycle, the solar flux number can easily go over 200.

Radio waves in your receiver

All radios, from a cheap AM radio to expensive communications receivers, have the same basic job to do. They have to take the incoming signals from the antenna, sep-

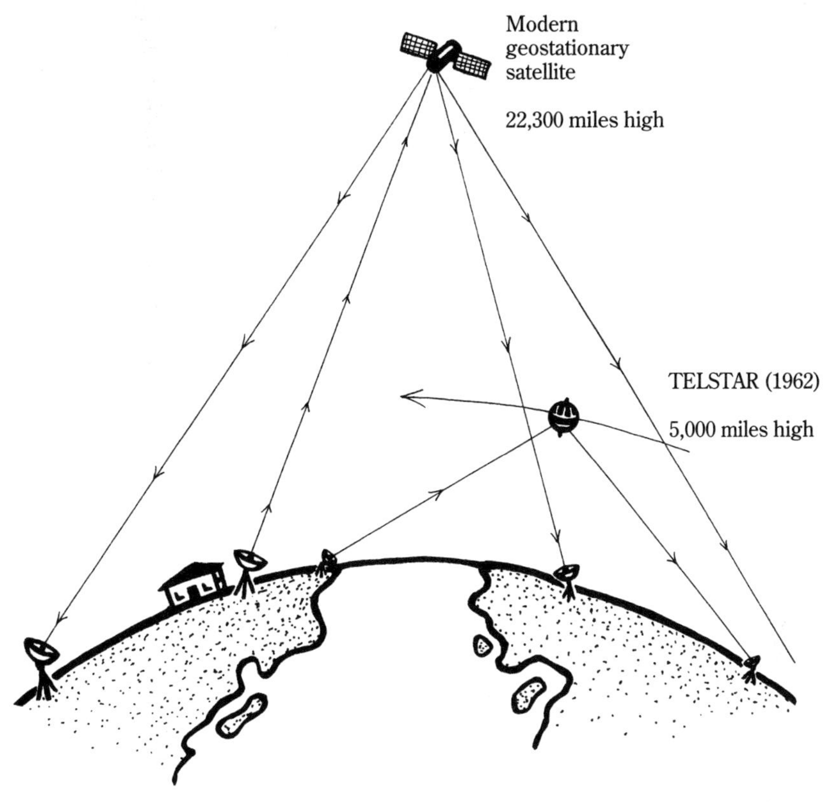

8-8 When radiation from a large solar flare hits the earth's magnetosphere, radio communication is disrupted.

arate the one you want to hear from all the others, amplify it, and change it from an electromagnetic wave into sound. Each job is done in a different stage of the radio, which is made up of a number of electronic parts or components. This is a brief rundown of how it works:

All radios have some kind of antenna system. Even if you can't see the antenna from the outside of the case, it's there. The antenna's job is to collect radio waves from space and bring them into the radio receiver.

When you put up an antenna, it collects all kinds of radio waves. Some of them you want to hear, and some you don't. Imagine what it would be like to turn on a radio and hear broadcast stations, police calls, TV shows, and ham radio operators—all on top of each other. You couldn't understand anything they were saying!

To prevent this from happening, your radio receiver has a tuner. Its purpose is to help you select the frequency of the signals you want to hear and cancel out all others.

Once the radio wave has left the tuner, it goes into the detector stage. Its job is to change the electromagnetic wave into an easier-to-manage electrical impulse. Then the signal comes into the reproducer, whose job is to change the electrical current coming from the tuner into sound. If you have a pair of earphones, you can now hear your program.

But if you want to use a speaker, you'll need a radio with an amplifier stage. As the name suggests, it amplifies the signal so you can enjoy the show without having to be attached to the radio. If the amplifier is powerful enough, you can hear the broadcast across the room, as you can with a small pocket radio. Or if you have a "boom box," the whole neighborhood will be able to hear the station you have selected.

Satellites for worldwide communication

Ever since the first communications satellites were launched into space in the early 1960s, they have been a vital part of TV and radio communications. The first communications satellite, Echo 1, was put into orbit during the summer of 1960. It was a 100-foot balloon, coated with a layer of aluminum designed to reflect radio waves.

AT&T's Telstar satellite, launched by NASA on July 10, 1962, was the first "active" communications satellite—meaning that it not only reflected radio waves as Echo 1 had done, but it could amplify and rebroadcast the signals that reached it. Telstar reached an altitude of 5,000 miles and carried the first TV broadcast between the United States and Europe (Fig. 8-9).

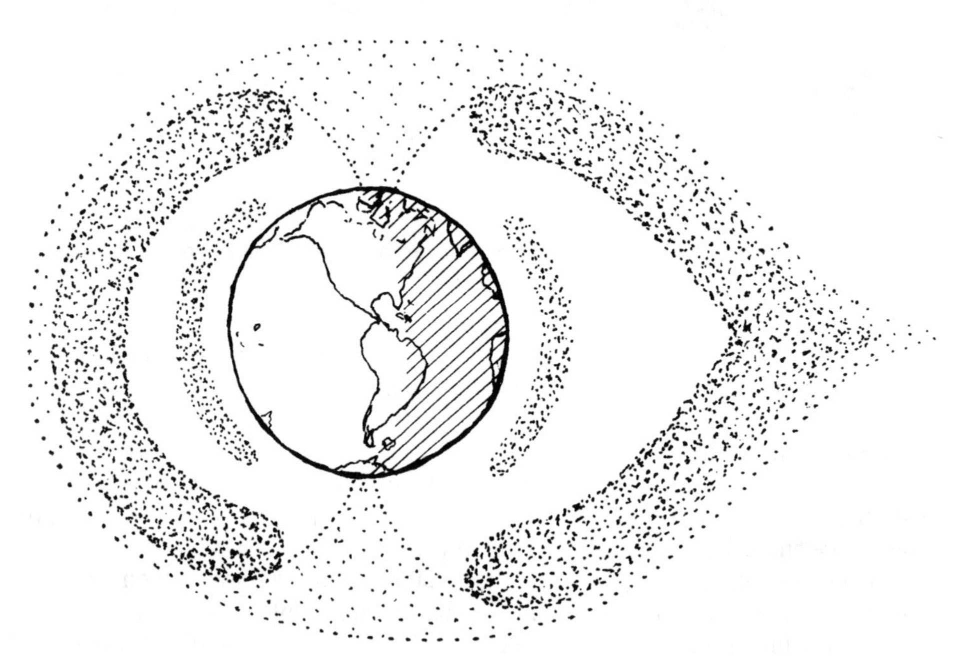

8-9 Today's communications satellites can broadcast over much greater distances than their early counterparts.

A few years later, a Syncom communications satellite was rocketed into orbit 22,300 miles above earth. This was a very important milestone, because a satellite orbiting our planet at 22,300 miles travels at the same speed as the earth's rotation. In effect, it holds the same position in the sky day and night. Once receiving and transmitting dishes are properly aimed at a geostationary satellite, they are able to do their job without having to be constantly readjusted.

A satellite in geostationary orbit above North America is high enough to cover most of the United States and the populated part of Canada with its "footprint." Three satellites, placed in geostationary orbit, can cover most of the world (Fig. 8-10).

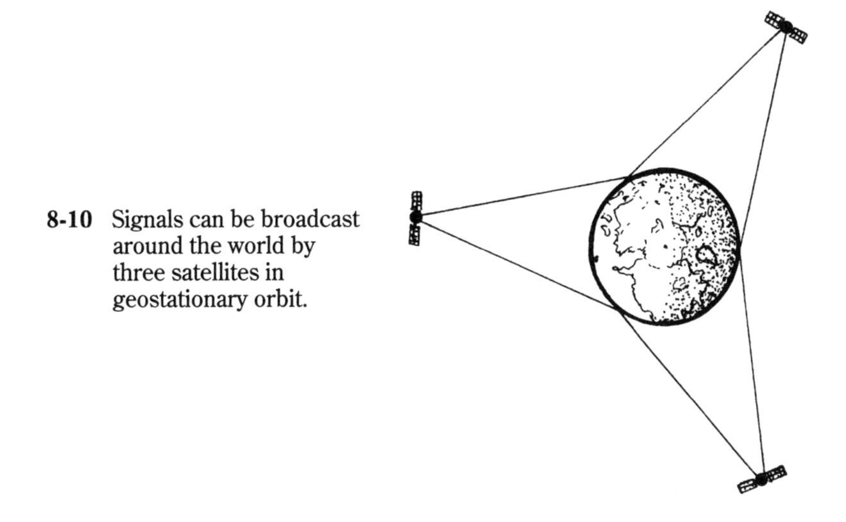

8-10 Signals can be broadcast around the world by three satellites in geostationary orbit.

Satellites in international broadcasting

International shortwave broadcasters use satellites to transmit programming to relay stations in other parts of the world. And an increasingly large number of shortwave stations, especially Europe, make their programs available on satellite for direct home-dish reception. Even when you're tuned to a local radio station or watching television, much of what you see and hear comes to you via satellite networks.

Radio and TV networks make extensive use of satellite communication to deliver programming to their affiliate stations. With the aid of satellite technology, a program can be produced in one city, uplinked to a satellite and downlinked to hundreds of stations throughout the country. Programming is then rebroadcasted on AM, FM, shortwave, and TV frequencies (Fig. 8-11).

Figure 8-12 shows Deutsche Welle's worldwide network of relay stations.

A satellite station in your backyard

In the past several years, more and more people have become involved in direct satellite reception. For less than a thousand dollars, you can have a complete satellite system in your own backyard—and tap into the space age way of picking up radio and television transmissions.

All network programming you see on your local TV channels is brought to you by satellite. It is picked up by a receiving dish at the station and retransmitted on fre-

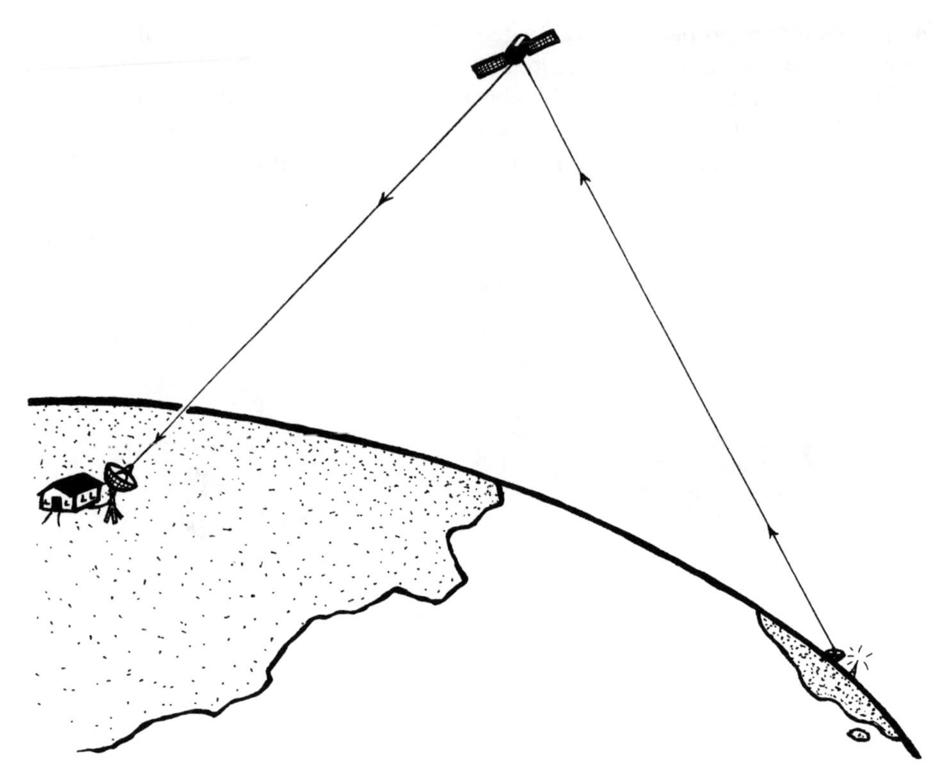

8-11 Satellite technology is used to transport programs across the country—and around the world.

quencies your television is capable of receiving. Cable companies also pick up most of their programming from satellites. And radio stations relay network news, talk shows, and music programs from satellites.

With a satellite receiving station in your backyard, you can receive all this (except for certain scrambled cable stations)—and much more. There are currently about 125 satellite TV stations available for dish owners to select from, as well as numerous news services, radio talk shows, music services, and so on.

The largest part of the system is the dish. Most satellite dishes are about 10 feet across. Their job is to collect signals coming from the satellite and direct them into the feed horn—which, in turn, brings them into an amplifier.

NOTE: Once a dish has been aimed at a geostationary satellite, it isn't necessary to readjust it—unless you want the dish to be in contact with a satellite orbiting over a different location. If you live in the United States or Canada, you'll need to aim your dish to the southeast or southwest to pick up most satellite transmissions.

The amplifier, also part of the dish setup, increases the strength of the signal about 10,000 times and changes the frequency from around 4 GHz down to 950–1450 MHz, which is an easier frequency for the receiver to handle.

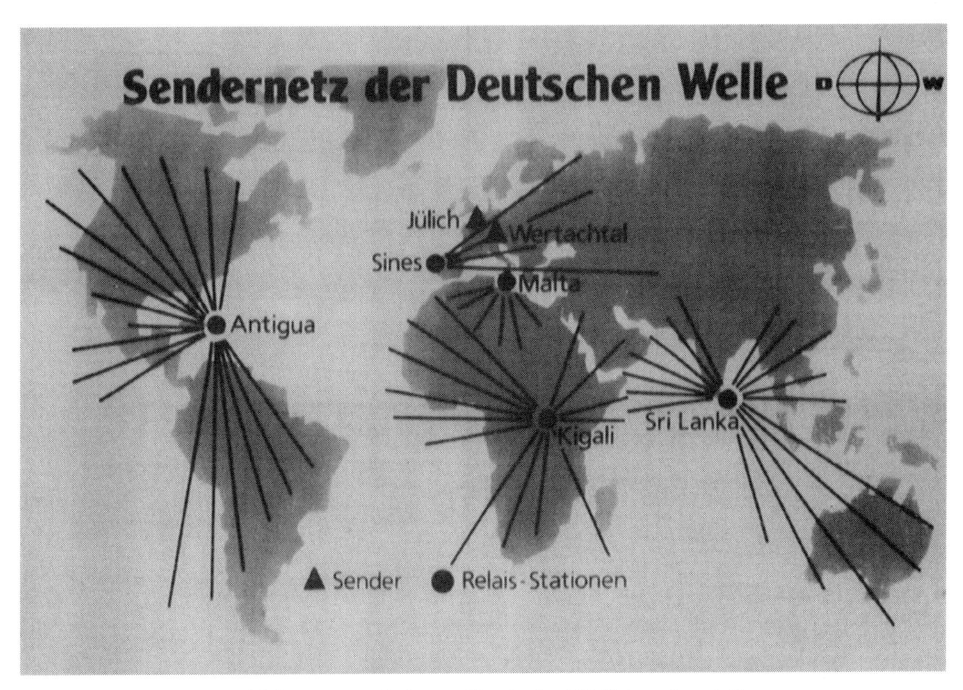

8-12 This map card shows Deutsche Welle's relay stations.

When the signal has been amplified and the frequency converted, it travels through a cable into the receiver—a rectangular unit you have in the house sitting next to the television. After the receiver has done its job, it sends the signal through another short cable into your TV. When it's all hooked up, you can sit back and enjoy hundreds of entertainment, news, music, and education services!

Appendix A
Shortwave radio stations

THIS APPENDIX CONTAINS THE ADDRESSES OF THE MOST COMMONLY HEARD shortwave stations that broadcast at least some of their programming in English.

Radio Tirana
Tirana
ALBANIA

Radio Argentina al Exterior—RAE
Casilla de Correo 555
1000 Buenos Aires
ARGENTINA

Radio Australia
GPO Box 428G
Melbourne VIC 3001
AUSTRALIA

Radio Austria International
A-1136
Vienna
AUSTRIA

Belgische Radio TV
Postbus 26
B-1000 Brussels
BELGIUM

Radio Nacional
C/P/ 04/0340
70323 Brasilia
BRAZIL

Radio Sofia
4 Dragan Tsankov Blvd
Sofia
BULGARIA

CFRX/CFRB
Ontario DX Association
PO Box 161, Station A
Willowdale, Ontario M2N 5S8
CANADA

Radio Canada International
PO Box 6000
Montreal PQ
H3C 3A8 Canada
CANADA

Radio CHU (Time & Frequency)
National Research Council
Ottawa, ON K1A OR6
CANADA

Radio Beijing
Beijing 100866
PEOPLE'S REPUBLIC OF CHINA

Voice of Free China
PO Box 24–38
Taipei, Tiawan
REPUBLIC OF CHINA

Radio for Peace International
PO Box 88
Santa Ana
COSTA RICA

Radio Habana Cuba
PO Box 70–26
Havana
CUBA

Radio Czechoslovakia
Czechoslovak Radio
12099 Prague 2
CZECHOSLOVAKIA

HCJB
Casilla de Correo 691 Quito
ECUADOR

Radio Cairo
Box 1186
Cairo
EGYPT

Radio Estonia
21 Lomonossovi
200 100 Talinn
ESTONIA

Radio Finland
Oy. Yelisradio Ab.
PL 95
SF-00241 Helsinki
FINLAND

Radio France International
B.P. 9516
F-75016 Paris Cedex 16
FRANCE

Deutsche Welle
Postfach 10 04 44
D-5000 Cologne 1
GERMANY

Radio Nederland
PO Box 222
Hilversum
HOLLAND

Radio Budapest
PO Box 1
H-1800 Budapest
HUNGARY

All India Radio—External Services
PO Box No. 500
New Delhi
INDIA

KOL Israel
PO Box 1082
91010 Jerusalem
ISRAEL

RAI Radiotelevisione Italiana
Viale Mazzini 14
00195 Roma
ITALY

Radio Japan/NHK
2–2–1 Jinnan
Shibuyaku, Tokyo
JAPAN

Radio Korea
18 Yoido-dong
Youngdungpo-gu
Seoul 150–790
REPUBLIC OF KOREA

Radio Vilnius
Konarskio 49
LT-2674
Vilnius MTP
LITHUANIA

Radio Luxembourg
B.P. 1002
Villa Louvigny à Luxembourg
LUXEMBOURG

Voice of the Mediterranean
Box 143
Valletta
MALTA

Trans-World Radio
Bonaire
NETHERLANDS ANTILLES

Radio New Zealand
PO Box 2092
Wellington
NEW ZEALAND

Voice of Nigeria
P.M.B. 12504
Ikoyi, Lagos
NIGERIA

Radio Norway International
0340 Oslo 3
NORWAY

Far East Broadcasting Co.
Box 1
Valenzuela, Metro-Manila 1405
PHILIPPINES

Polish Radio, Warsaw
00 950 Warsaw
POLAND

Radio Portugal International
Rua Sao Marcal 1
Lisbon
PORTUGAL

Radio Romania International
Str. Nuferilor 60–62
79756 Bucharest
ROMANIA

Radio Moscow
25 Pyanitskaya Street
Moscow 113326
RUSSIA

Radio RSA
Piet Meyer Building
Henley Road
Broadcasting Centre
Johannesburg 2000
REPUBLIC OF SOUTH AFRICA

Spanish Foreign Radio
Apartado de Correo 156.202
E-28080 Madrid
SPAIN

Radio Sweden
S-105 10 Stockholm
SWEDEN

Swiss Radio International
PO Box CH3000
Berne 15
SWITZERLAND

Turkish Radio TV Corporation
P.K. 333
Ankara
TURKEY

Radio Kiev
Kiev
UKRAINE
BBC World Service
PO Box 76
Bush House, London WC2B 4PH
UNITED KINGDOM

Christian Science Monitor
Shortwave World Service
1 Norway Street
Boston, MA 02115
USA

KGEI–Voice of Friendship
1400 Radio Road
Redwood City, CA 94065
USA

KNLS
PO Box 473
Anchor Point, AK 99556
USA

KVOH
High Adventures Ministries
Box 7466
Van Nuys, CA 91409
USA

RadioNewyork International (via WWCR)
507 Violet Avenue
Hyde Park, NY 12538
USA

WHRI
World Harvest Radio
PO Box 12
South Bend, IN 46624
USA

WINB
PO Box 88
Red Lion, PA 17356
USA

WRNO
4539 I-10 Service Road N.
Metairie, LA 70002
USA

WWCR
1300 WWCR Ave.
Nashville, TN 37218
USA

WWV (Time & Frequency)
2000 East Coast Rd. 58
Ft. Collins, CO 80524
USA

WWVH (Time & Frequency)
PO Box 417
Kekaha, Kauai, HI 96752
USA

WYFR
Family Radio
290 Hegenberger Road
Oakland, CA 94621
USA

Vatican Radio
Vatican City
VATICAN CITY

Radio Yugoslavia
PO Box 2000
11000 Belgrade
YUGOSLAVIA

Appendix B
Radio listening clubs

JOINING A RADIO LISTENING CLUB IS ONE OF THE BEST WAYS TO KEEP UP WITH current information on frequency changes, programming, and so forth. Most clubs are run by volunteers who have been enjoying their listening hobby for some time and would like to share their knowledge with others.

Organizations of radio clubs

Each of the following organizations serves many clubs and can supply you with a more complete and detailed listing of radio clubs in the area it serves. Enclose return postage or International Reply Coupons with your letter to ensure a quick response.

ANARC (The Association of North American Radio Clubs)
2216 Burkley Drive.
Wyomissing, PA 19610
USA

EDXC (European DX Council)
PO Box 4
St. Ives
Huntingdon, Cambs. PE17 4FE
ENGLAND

Association of Pan-Asian Radio Clubs
PO Chabdana-1702
Dt. Gazipur
BANGLADESH

South Pacific Association of Radio Clubs
c/o NZ Radio DX League
212 Earn Street
Invercargill
NEW ZEALAND

National and international clubs

If you would like to receive a sample bulletin from any of the foreign groups, send U.S. $2 or three International Reply Coupons.

North American DX clubs

American Radio Relay League
225 Main Street
Newington, CT 06111
203–666–1541
(amateur radio)

American Shortwave Listeners' Club
Stewart MacKenzie, WDX6AA
16182 Ballad Lane
Huntington Beach, CA 92649
714–846–1685
(shortwave)

*Association of Clandestine Enthusiasts (The A*C*E*)
Kirk Baxter
PO Box 11201
Shawnee Mission, KS 66207
(pirates and clandestine)

*Association of DX Reporters
Reuben Dagold
7008 Plymouth Road
Baltimore, MD 21208
(shortwave, ham, utility, AM, and longwave)

Bearcat Radio Club
PO Box 291918
Kettering, OH 45429
1–800–423–1331
(VHF/UHF scanning)

*Canadian International DX Club
Sheldon Harvey, President
79 Kipps Street Road
Greenfield Park, Quebec T8G 1A5
CANADA
514–462–1459
(all bands)*

Longwave Club of America
45 Wildflower Road
Levittown, PA 19057
(longwave band—below 550 kHz)

* Miami Valley DX Club
Box 291232
Columbus, OH 43229
(shortwave)

National Amateur Radio Association
PO Box 598
Redmond, WA 98073–0598
1–800–468–2426
(amateur radio)

* National Radio Club
PO Box 5711, Dept. ALM
Topeka, KS 66605–0711
(AM)

* North American Shortwave Association
Bob Brown, Executive Director
45 Wildflower Road
Levittown, PA 19057
(shortwave)

* Radio Communications Monitoring Association
PO Box 542
Silverado, CA 92676
(two-way communication, UHF/VHF scanning)

* SPEEDEX
(Society to Preserve the Engrossing Enjoyment of DXing)
Bob Thunberg, Business Manager
PO Box 196
DuBois, PA 15701–0196
(shortwave broadcasts and utilities)

Worldwide TV-FM DX Association
PO Box 514
Buffalo, NY 14205
(FM, VHF/UHF, satellites, and TV)

DX clubs in Africa, Asia, Australia, Europe, and Latin America

African DX Association
c/o Friday Okoloise
Radio/Carrier Room
Nitel, Ashaka
Bauchi State
NIGERIA

ANDEX
Radio HCJB
PO Box 619
Quito
ECUADOR
(shortwave)

Australian DX Club
Box 227
Box Hill, Vic. 3128
AUSTRALIA
(shortwave)

British DX Club
54 Birkhall Road
Catford, London SE6 1TE
ENGLAND
(shortwave)

Czechoslovak DX Club
c/o Vasclav Dosoudil
Horni 9
CS-76821
Kvasice
CZECHOSLOVAKIA
(shortwave)

Danish Shortwave Clubs International
Taveager 31
DK-2670 Greve
DENMARK
(shortwave)

International Listeners' Association
1 Jersey Street
Hafod, Swansea SA1 2HF
ENGLAND
(shortwave)

International Shortwave League
6 Moorhead
Preston Upon The Weald Moors
Telford, Shropshire TF6 6DC
ENGLAND
(shortwave and ham radio)

Pakistan SW Listeners' Clubs Association
Javaid Iqbal
PO Box 5
Sheikhupura 39359
PAKISTAN
(shortwave)

Radio Budapest Shortwave Club
Radio Budapest
PO Box 1
H-1800 Budapest
HUNGARY
(quarterly paper mailed free to R. Budapest listeners who request it)

South African DX Club
PO Box 72620
Lynwood Ridge
Transvaal 0040
SOUTH AFRICA
(shortwave)

Regional clubs

The following clubs draw in listeners from the clubs' local areas. Some focus on VHF/UHF (scanner) frequency information, which would be of little or no use to people living elsewhere. And some clubs prefer to stay local so that members can meet face-to-face to discuss their latest catches, show off QSL cards, equipment, and so on. Most publish some kind of bulletin or newsletter and will be happy to send you a sample copy for return postage.

All Ohio Scanner Club
Dave Marshall
50 Villa Road
Springfield, OH 45503–1036
(VHF/UHF and ham)

Association of Manitoba DXers
Shawn Axelrod
30 Becontree Bay
Winnipeg, Manitoba, R2N 2X9
CANADA
204–253–8644
(AM, shortwave, and VHF/UHF)

Bay Area Scanner Enthusiasts
4718 Heridian Avenue #265
1465 Porobelo Drive
San Jose, CA 95118
(UHF/VHF scanning)

Boston Area DXers
Paul Graveline
9 Stirling Street
Andover, MA 10810
508–470–1971
(shortwave)

Capitol Hill Monitors
Alan Henney
6912 Prince Georges Avenue
Takoma Park, MD 20912
(UHF/VHF scanning)

Chicago Area Radio Monitoring Association
Kurt Stoudt
2625 North Forest
Arlington Heights, IL 60004
(UHF/VHF scanning)

Chicago DX Club
c/o Thomas V. Ross
8225 West 43 Place
Lyons, IL 60534
(mostly shortwave)

Cincinnati Area Monitoring Exchange
John Vodenik
513–398–5968
(all bands)

DX South Florida
3156 NW 39 Street
Ft. Lauderdale, FL 33309
(shortwave)

Frequency Fan Club
Race Scanning Monthly
PO Box 991
Mulberry, FL 33860
(covers auto race scanner frequencies)

Hawkeye Scanning Group of Iowa
PO Box 974-HS
Burlington, IA 52601–0974
(VHF/UHF)

Metro Radio System
Julian Olanskay
PO Box 26
Newton Highlands, MA 02161
617–969–3000
(New England public safety frequencies)

Michigan Area Radio Enthusiasts
Bob Walker
PO Box 311
Wixom, MI 48393
(all bands)

Minnesota DX Club
PO Box 3164
Burnsville, MN 55337
(all bands)

* Monitor Communications Group
Louis Campagna
8001 Castor Avenue #143
Philadelphia, PA 19152
(all bands)

Northeast Ohio DXers
Mike Fanderys
2802 North Avenue
Parma, OH 44134
216–661–2443
(shortwave and utility stations)

NE Ohio SW Listeners
Brian Schaft
317 South Rocky River Drive
Berea, OH 44071
216–234–4628
(shortwave)

Northeast Scanner Club
Les Mattson
PO Box 62
Gibbstown, NJ 08027
(UHF/VHF scanning, Maine through Virginia)

Ontario DX Association
Harold Sellers
PO Box 161—Station A
Willowdale, Ontario M2N 5S8
CANADA
(all bands)

Radio Monitors Newsletter of Maryland
Ron Bruckman
PO Box 394
Hampstead, MD 21074
(UHF/VHF scanning and shortwave)

Regional Communications Network
Bill Morris
Box 83-M
Carlstadt, NJ 07072–0083
(all bands within 50-mile radius of New York City)

Rocky Mountain Radio Listeners
Wayne Heinen
4131 S. Andes Way
Aurora, CO 80013–3831
(all bands in Denver metropolitan area)

Scanning Wisconsin
c/o AJC, Inc.
W. 17912 Pearl Drive
Muskego, WI 53150
(VHF/UHF scanning)

* Southern California Area DXers
Don R. Schmidt
3809 Rose Avenue
Long Beach, CA 90807–4334
310–424–4634
(all bands)

Toledo Area Radio Enthusiasts
Ernie Dellinger, N8PFA
6629 Sue Lane
Maumee, OH 43537
419–865–4284
(all bands)

Virginia Monitoring Digest
PO Box 34832
Richmond, VA 23234
(all bands)

* Washington Area DX Association
606 Forest Glen Road
Silver Spring, MD 20876
(all bands)

---

* ANARC members or associate members

Appendix C
Radio sources

THIS APPENDIX CONTAINS THE NAMES OF COMPANIES THAT SUPPLY THE RADIO receivers, transmitters, accessories, antennas, and books you'll need to get the most out of your new hobby.

ACE Communications
Monitor Division
10707 East 106th Street
Fishers, IN 46038
1–800–445–7717

(VHF/UHF scanners.)

ACME Enterprises
1358 Coney Island Avenue
Suite 200 Z
Brooklyn, NY 11230

(Books on pirate radio.)

American Radio Relay League
225 Main Street
Newington, CT 06111

(*QST Magazine*, ham radio books, ham license study guides.)

Antenna Supermarket
PO Box 563
Palatine, IL 60078

(Multiband shortwave antennas.)

Antennas West
1500 North 150 West
Provo,UT84605
(Shortwave ham and scanner antennas.)

Austin Amateur Radio Supply
5325 North IH-35
Austin, TX 78723

(Ham radio transceivers, accessories, and repair service.)

Berry Electronics Supply Company
512 Broadway
New York, NY 10012

(Ham radio transceivers, accessories, and repair service.)

Chilton Pacific Ltd.
5632 Van Nuys Boulevard #222
Van Nuys, CA 91401

(High-performance, long-range GE Superadio II—AM/FM portable.)

Communications Electronics Inc.
PO Box 1045-PC 92
Ann Arbour, MI 48106

(Ham, shortwave, and CB equipment and service.)

CQ Communications
76 North Broadway
Hicksville, NY 11801

(*CQ* ham radio magazine, *CQ Amateur Radio Equipment Buyer's Guide, CQ Antenna Buyer's Guide,* ham radio books.)

CRB Research Books
PO Box 56
Commack, NY 11725

(Communications books, frequency guides for all bands.)

Drake Company, R. L.
PO Box 3006
Miamisburg, OH 45324
1–800–937–2534

(Shortwave radios.)

DX Radio Supply
PO Box 360
Wagontown, PA 19376
215–273–7823

(*National Scanning Report Magazine, Betty Bearcat Scanner Guides,* and other communications books.)

Electronic Equipment Bank
323 Mill Street, NW
Vienna, VA 22180
1–800–368–3270

(Ham radio transceivers, shortwave receivers, and repair service.)

Galaxy Electronics
67 Eber Avenue
PO Box 1201
Akron, OH 44309
216–376–2402

(New and used shortwave radios and scanners.)

Gilfer Shortwave
52 Park Avenue
Park Ridge, NJ 07656

(Shortwave radios and communications books.)

Gordon West Radio School
PO Box 2013
Lakewood, NJ 08701

(Ham radio license study guides, code practice cassettes, and classroom study supplies.)

Grove Enterprises, Inc.
140 Dog Branch Road
PO Box 98
Brasstown, NC 28902
1–800–438–8155

(*Monitoring Times* magazine, shortwave radios, scanners, books, and antennas.)

Grundig
3520 Haven Avenue
Unit L
Redwood City, CA 94063

(Shortwave radios.)

Hamstuff
PO Box 14455
Scottsdale, AZ 85267

(QSL storage boxes, ham radio T-shirts, etc.)

Hamtronics, Inc.
4033 Brownsville Road
Trevosee, PA 19047
1–800–426–2820

(Ham radio equipment.)

HR Bookstore
PO Box 209
Rindge, NH 03416
1–800–457–7373

(Ham and shortwave radio books.)

Hustler Antennas
One Newtronics Place
Mineral Wells, TX 76067

(Ham, scanner, and CB antennas.)

ICOM America, Inc.
2380 116th Avenue, NE
Bellevue, WA 76067

(Ham radio transceivers and accessories.)

Japan Radio Çompany, Ltd.
430 Park Avenue, 2nd Floor
New York, NY 10022

(Shortwave receivers, ham transmitters, and accessories.)

Kenwood USA Corporation
2201 East Dominguez Street
PO Box 22745
Long Beach, CA 90801

(Shortwave receivers, ham radio transceivers, and accessories.)

Lentini Communications, Inc.
21 Garfield Street
Newington, CT 06111
1–800–666–0908

(Shortwave radios, scanners, and ham radio equipment.)

Midland International Corporation
1690 North Topping Avenue
Kansas City, MO 64120

(CB radios, VHF marine radios, antennas, and accessories

Mil-Spec Communications
PO Box 461
Wakefield, RI 02880
401–783–7106

(Shortwave radios and repair service)

National Amateur Radio Association
PO Box 598
Redmond, WA 98073
1–800–468–2426

(*Amateur Radio Communicator* magazine, license study materials.)

National Scanning Report
PO Box 291918
Kittering, OH 45249
1–800–423–1331

(Publishes *National Scanning Report* magazine.)

Offshore Echoes
PO Box 1514
London W72LL
ENGLAND

(Catalog of tapes, records, CDs, posters, etc., of offshore broadcast stations of the past and present—including Radio Caroline and Radio Newyork International.

Palomar Engineers
PO Box 462222
Escondido, CA 92046
619–747–3343

(Antennas, preamplifiers, and filters.)

Philips
1–800–328–0795

(Call for dealer information on the Philips DC-777 AM/FM/Shortwave Cassette car stereo.)

PIF Books by Mail
PO Box 888
Hawkensbury, Ontario
Canada K6A 3E1

Popular Communications
76 North Broadway
Hicksville, NY 11801

(*Popular Communications* magazine and annual *Popular Communications Guide*.)

Popular Electronics
PO Box 338
Mt. Morris, IL 61054
1–800–435–0715

(*Popular Electronics* magazine.)

QSLs by W4MPY
682 Mount Pleasant Road
Monetta, SC 29105

(QSL cards and logbooks.)

Radio Amateur Callbook
PO Box 2013
Lakewood, NJ 08701

(Ham radio callbooks, radio maps, license study materials.)

Radio Buffs
1–800–828–6433

(Call for price quotes on ham and shortwave books and equipment.)

The Radio Collection
PO Box 149
Briarcliff Manor, NY 10510

(Shortwave radio books, souvenirs, and low-power FM transmitters.)

Satman, Inc.
6310 North University No. 3798
Peoria, IL 61612
1–800–472–8626

(Satellite TV equipment.)

Scanner World, USA
10 New Scotland Avenue
Albany, NY 12208
518–436–9606

(Scanners, shortwave radios, CBs, antennas, and radio books.)

Skyvision, Inc.
1050 Frontier Drive
Fergus Falls, MN 56537
1–800–334–6455
(Satellite TV receiving equipment.)

Somerset Electronics, Inc.
1290 Highway A1A
Satellite Beach, FL 32937
1–800–678–7388

(Multimode decoders for receiving digital transmissions from shortwave utility stations.)

TAB Books
Blue Ridge Summit, PA 17294-0214

(Shortwave, ham radio, and antenna books.)

Tiare Publications
PO Box 493
Lake Gineva, WI 53147

(Specializes in books on pirate and clandestine radio.)

Turbo Electronics
PO Box 8034
Hicksville, NY 11834
1–800–33–TURBO

(Scanners, CBs, and other electronic equipment.)

Universal Radio Inc.
6830 American Parkway
Reynoldsburg, OH 43068
1–800–431–3939

(Shortwave radios, ham radio, scanners, antennas, and books.)

W5YI Group
PO Box 565101
Dallas, TX 75356
1-800-669-9594

(Ham radio license study guides and code practice cassettes.)

Glossary

aeronautical radio Communications between aircraft and ground control, or between one aircraft and another.

amateur call letters Ham radio identification letters. In the USA, they are issued by the Federal Communications Commission.

antenna Wire or other metal device used to gather radio waves and direct them into a receiver.

ARRL (American Radio Relay League) The largest ham radio organization in the United States. Publishes *QST* magazine.

ASCII (American Standard Code for Information Interchange) Most commonly used code for exchanging information via computer.

bandwidth Amount of frequency space taken up by a signal.

beam antenna A directional antenna used by many hams.

BFO (beat frequency oscillator) A feature on some receivers that makes single sideband voice transmissions intelligible, and improves Morse code reception.

clandestine station Station broadcasting political messages without a license, usually in connection with revolutionary movements.

digital readout Display on a radio that shows exactly which frequency you are on.

domestic station Station that broadcasts programs intended for a local or regional audience.

downlink Frequency used to relay satellite transmissions to earth.

DX Distance.

electromagnetic waves Energy waves made up of an electrical and magnetic field, used to transmit radio and TV signals.

feed horn Part of a satellite dish antenna that collects the reflected signal and funnels it into the amplifier.

footprint Area covered by satellite transmissions.

frequency Number of radio waves that pass a given point per second.

geomagnetic storms Disruption of earth's magnetosphere caused by solar flares.

geostationary orbit An orbit 22,300 miles above us where satellites circle our planet at the same speed as the earth's rotation.

GHz Million hertz.

GMT (Greenwich mean time) Old term for UTC, the worldwide standard time zone used in shortwave radio.

ground wave Radio wave that stays near the earth, and can be heard only for a limited distance.

HF (high frequency) 3 to 30 MHz (includes shortwave bands).

IRC (international reply coupon) Worldwide exchange medium used to pay for postage costs.

ionosphere Layers of electrically charged gas in our atmosphere that affect radio wave skip.

ITU (International Telecommunications Union) International organization that regulates use of the radio wave spectrum.

ITU Phonetics Worldwide system of phonetics, recommended by the ITU for use in voice communications when reception conditions are difficult.

kHz (kilo hertz) Thousand hertz, or radio waves, per second.

line of sight Straight, noncurving path radio waves take on VHF and UHF frequencies.

LSB (lower sideband) Single sideband transmission with the upper sideband and carrier removed.

magnetosphere Magnetic field that surrounds the earth and protects us from much of the sun's harmful radiation.

medium wave Frequencies used for AM band broadcasts.

MHz (mega hertz) Million hertz, or radio waves per second.

mode Method of transmitting, such as AM, FM, Morse code, and radioteletype.

MUF (maximum usable frequency) The highest frequency you can use to transmit from one area to another.

multimode decoder Device used to interpret digital (CW, RTTY, ASCII, and other modes) transmissions so you can view them on a TV or computer screen.

pirate station Hobby broadcast station, operated without a license.

propagation Transportation of radio waves through the atmosphere from one part of the world to another.

QRP Low-power operation (in ham radio, usually under 10 watts).

Q-signals Ham radio abbreviations for frequently used messages.

QSL card A card from a radio station that verifies that you heard their broadcast.

reception report Report sent to a station to request a QSL card.

relay station A station that picks a signal up on one frequency (usually from a satellite downlink) and retransmits it on another (AM, FM, or shortwave).

repeater Automated transmitter used on high-frequency ham bands to pick up, amplify, and retransmit a signal.

RST (readability, signal, tone) Signal reporting system used by ham radio operators.

RTTY (radioteletype) Mode of digital transmission that is much faster than Morse code and is used by hams and utility stations.

selectivity The ability of a receiver to separate one station from another nearby broadcaster.

sensitivity The ability of a receiver to amplify weak signals so you can better hear them.

shortwave Radio frequencies between 3 MHz and 30 MHz.

SINPO (signal, interference, propagation, overall) Signal reporting system used by shortwave listeners.

skip zone The area where you are too far away from the station to hear the ground wave, but too close to receive a skip.

solar flare A powerful explosion on the sun that can cause blackouts in shortwave radio reception here on earth.

solar flux index A method of measuring the solar activity that influences radio wave propagation.

SSB (single sideband) Mode of transmission used by ham radio operators, utility stations, and a few shortwave stations. One sideband is cut off to save on bandspace—and must be replaced by using the SSB/BFO control on your receiver.

UHF (ultra high frequency) 300 MHz to 3 GHz.

uplink Frequency used to transmit signals from an earth station up to a satellite.

USB (upper sideband) Single sideband transmission with the lower sideband removed.

utility station A wide assortment of stations that don't broadcast to the general public and aren't ham radio operators. Aviation, ship-to-shore, international weather and military transmissions are included in this category.

VHF (very high frequency) 30 MHz to 300 MHz.

wavelength The distance from the peak of one radio wave to the peak of the next.

WPM (words per minute) Number of words sent per minute (Morse code).

XCVR (transceiver) Ham receiver and transmitter in one unit.

Index